MUSIC OF THE EARTH

Volcanoes, Earthquakes, and Other Geological Wonders

MUSIC OF THE EARTH

Volcanoes, Earthquakes, and Other Geological Wonders

RON L. MORTON

PLENUM PRESS • NEW YORK AND LONDON

Library of Congress Cataloging-in-Publication Data

Morton, R. L. (Ronald Lee)
Music of the earth : volcanoes, earthquakes, and other geological wonders / Ron L. Morton.
p. cm.
Includes bibliographical references and index.
ISBN 0-306-45263-4
1. Earth sciences--Popular works. I. Title.
QE31.M67 1996
550--dc20
96-1764
CIP

ISBN 0-306-45263-4

Plenum Press is a Division of Plenum Publishing Corporation
233 Spring Street, New York, N.Y. 10013-1578

10 9 8 7 6 5 4 3 2 1

Printed in the United States of America

Acknowledgments

When you start you never know
All the strange places you may go.

These two lines are from a poem about a river, but they could just as easily apply to the places where books are born—you just never know.

This book owes its form and content to the interest and enthusiasm of the great group of people who attended Elderhostel at Gunflint Lodge, the University of Minnesota at Duluth, in May of 1993 and 1994. Also at Gunflint, with her keyboard piano and wonderful stories was the musician and teacher Jean Swanson. Her presentations on "Nature in Music" were the beginnings of the idea for *Music of the Earth*.

Paul Schmitz graciously read the entire manuscript and, with his interest in geology and his great love of music, provided me with numerous ideas on combinations of instruments that would actually make the music created by the earth. Thanks Paul, I only hope the final choices meet with your musical approval.

To Megan and Chris I owe the debt of encouragement to actually write.

To Charlie Matsch and Penny Morton many thanks for reading early chapters and/or rough drafts of this manuscript, and for all the good comments and ideas for improvement.

Then there is Penny, to whom, very simply, I owe everything; and Froto and Elsie, whose time will come.

The editorial suggestions of Linda Regan of Plenum—her comments, deletions, and additions—were terrific (even when she put great, slashing blue lines through most of what she referred to as my cornball humor). Her work has greatly enhanced this book.

Finally, there is geology; I am continually in awe of this truly wonderful, amazing world we live on. I am also thankful for a science that has enabled me to see and experience a little of it in a very personal way. With geology, as with all else,

When you start you never know
All the strange places you may go.

Ron Morton

An invitation . . .

Here about the beach I wander
Nourishing a youth sublime
With the fairy tales of science
And the long results of time
—ALFRED, LORD TENNYSON

. . . if you're ready . . .

The most beautiful and most profound emotion we can experience is the sensation of the mystical. It is the sower of all true science. He to whom this emotion is a stranger, who can no longer wonder and stare rapt in awe, is as good as dead.
—ALBERT EINSTEIN

. . . to the greatest wonder of all.

A planet called earth.
A small world, our home,
A blue sphere of dynamic processes,
Processes that lead to motion, vitality, life,
Totally awesome.
And we humans
Are part of and contribute to this awesomeness,
The earth and us,
Spinning together through the "fairy tale" of time.

Contents

MUSIC OF THE EARTH

Volcanoes, Earthquakes, and Other Geological Wonders

Introduction

More Than the Flicker of an Eye

Music can be defined as sound arranged into pleasing or interesting patterns, especially as produced by voices or instruments. Longfellow called music "the universal language of mankind"; Caryle called it "the speech of angels." Not hard to imagine. Recall the long, haunting wail of Paul Winter's saxophone, the whizzes and hums of Yo-Yo Ma's cello, the deep richness of Pavarotti's voice, the rhythm and beat of Paul Simon's songs.

Feeling, pleasure, wonder—human voices and instruments making music.

Earth music may not be the voice of angels, but it does give us a rich, interesting pattern of sounds, sounds produced by the instruments of our dynamic earth: the long, wavering rumble of an earthquake, the beat and rhythmic roar of a volcano, the eerie cracks and snaps of glacial ice, the whistles and fizzes of hot springs.

Feeling, wonder, awe, and sometimes fear—earth instruments making earth music.

George Bernard Shaw called music "the brandy of the damned." On the same note we might call earth music "the tonic of the stranded." Tonic because earth music stirs, agitates, and shakes. It is music that keeps everyone on the planet awake and on our toes. Stranded because this old house, this earth, is our permanent home. It was our ancestors' home, and it will be our children's children's home. We can't move to a better neighborhood, get a bigger house with a better view—at least not in the foreseeable future. So we need to take a good look at exactly where we are and on what we are living. We need to find out what our home is made of, how everything works, and if it's possible to redecorate or make repairs. Finally, we need to remember Will Durant's admonition that "civilization exists by geological consent . . . subject to change without notice."

But can we consider the sounds, tones, and noises created by the instruments of the earth to be music? That's a question we all can ask and answer for ourselves, just as we do when we ask if the sounds, tones, and noises created by bagpipes, electric guitars, or tenor voices are music.

Music or noise, rhythm or jangles, melody or squeaky hinges. The decision rests with the ear of the listener, and, given individual tastes, that's right where it should stay.

But for a moment, imagine an earth symphony played by instruments made by us—the new kids in a rather old neighborhood.

A distant morning, desolate and steamy (horns and oboes, low and melodic). Sunrise over an empty ocean. The earth stretches; it wiggles and breathes. Ocean floors begin to move and continents start to wander (woodwinds and trumpets have found their voices). Migrating continents and sliding oceans cause volcanic eruptions and earthquakes (hear the power of the pipe organ and the low, steady moan of the double basses). Volcanos give birth to hot springs, geysers, and mineral waters (a lull for clarinets and flutes). These dynamics contribute to and help drive the earth's atmosphere, producing climatic change and ice ages (an interlude for French horns and cellos). This is earth music: earth rhythms from earth processes.

Much like the instruments of an orchestra, the instruments of the earth (earth processes) are interrelated and depend on each other to make the symphony work. These processes ebb and flow, over and over, again and again, eon upon eon. Continents wander, volcanoes roar, the earth rumbles, geysers spout and mud pots bubble, climates change, ice advances and retreats, species rise and fall.

What is often true in our lives is also true in the life of the earth—what goes around, comes around. Earth music has not changed all that much through the eons of geologic time. Sure, on occasion the pipe organ will be loud and boisterous, but it soon quiets and gives way to the clarinet and flute, which may be followed by the French horn and cello. But the symphony being played—the patterns of notes, the overall rhythm, the flowing melody—is similar across the corridors of geologic time.

Geologic time—we know the earth is very old, but it occurs to me that the millions, nay billions, of years of geologic time, which geologists constantly remind us of, and impress us with, are not what an understanding of earth history should really be about. The span of geologic time is almost impossible to equate with the daily life of any of us. It is like our budget deficit—difficult to imagine, hard to comprehend, and seemingly impossible to control. Geologic time is simply not relevant to most of us.

Yes, the earth is very old and its oldness can make one feel rather humble. If all of geologic time were compressed into one calendar year, we who walk erect would arrive on center stage at 4 p.m. on December 31, and the Roman Empire would last for 5 seconds! Now aren't we special.

But the processes of the earth go on, over and over. If we can see and describe these processes—these instruments of time—if we can understand how they work, then we can begin to know our geologic past and start to taste our future. The vastness of geologic time becomes like a corkscrew turning yesterday's sunset into tomorrow's sunrise. With this understanding we just might remain part of the orchestra and thus become more than the flicker of an eye.

So fasten your seatbelts and come along for a ride on the corkscrew of time. Together, with open eyes and minds, with curiosity, imagination, and humor, we will explore the instruments that make the music of the earth.

Prelude to Chapter 1

Eon after eon continents creep and ocean floors slide across the planet's surface, mostly to the sounds of silence. Occasionally, however, two continents collide, crashing and smashing together, or an ocean floor cracks and breaks and molten rock bubbles to the surface. Then, a slow return, mostly to the sounds of silence.

Imagine long-held chords of the woodwinds, soft and slow as the eons flow by; then, a sudden fanfare bursts from the trumpets as two continents meet or an ocean floor breaks apart. The fanfare fades, and the woodwinds slowly return, long, low, and soft, as the continents and ocean floors continue on their great, silent journey.

CHAPTER 1

Those Creepy Continents

Early Greek mathematicians, including Pythagoras of theorem fame, believed that the movement of the planets produced beautiful, harmonious music. The Greeks called it "music of the spheres," but unfortunately we who live on earth are unable to hear it. The fact is that we are not perfect enough, not gods, and this celestial music is only for the ears of the gods.

The earth's part in this planetary symphony may well be played out deep inside the planet. Down, far from the maddening surface, down hundreds of miles where great forces toil and struggle to move ocean floors and continents. The sounds of those struggles, the noises, tones, and notes, may well be beyond the mortal realm; it may be part of the "music of the spheres."

Music we can't hear playing as oceans spread and continents creep—that's enough to send shivers through anyone. Creeping continents—do you believe it? You certainly can't see them move. We all know clouds drift across the sky and cars move. Aunt Clara moved east and Fred went west, but how about the house, the yard, the city, or town? It isn't going anywhere—is it?

Well, first the bad news. It sure is. Continents do creep. New York continues to get farther and farther from London (that's why airfares keep going up). The good news is that continents only creep. They don't walk, run, or fly. By creeping along at a snail's pace, actually at a speed about equal to the growth of your fingernails, continents give the false impression of being permanent, forever fixed in one spot.

You can sit and stare at your fingernails for hours on end and they don't seem to grow. But wait two or three weeks and all of a sudden you're running around trying to find the nail scissors. Fingernails and continents—only instead of weeks, a person living in Florida would have to wait millions and millions of years before waking up to the snow of Minnesota.

FROM BACON TO WEGENER

Moving continents—Egad!—Galileo may have guessed as much when he said, "But it does move!" (Of course, we really don't know if he was referring to the European continent or to his belief that the planet Earth traveled around the sun.) Francis Bacon certainly suspected. The English philosopher proposed the idea of moving continents some 370 years ago (let's see, your fingernail would be about 6 feet [1.8 m] long—that's how far New York has moved away from London). Bacon had access to the early maps that had been made of the coastlines of South America and Africa. Upon seeing these he commented that they appeared to match and wondered if one was torn from the other.

In 1855 Antonio Snider published a sketch map showing South America and Africa joined together. Fitting these two continents together in this jigsawlike manner gave rise to the bold and daring suggestion that perhaps they had once been part of the same landmass, a landmass that had somehow been broken and forced to drift apart. The scientific community shook their collective heads at such a preposterous idea. Continents break and move? Humbug and tommyrot!

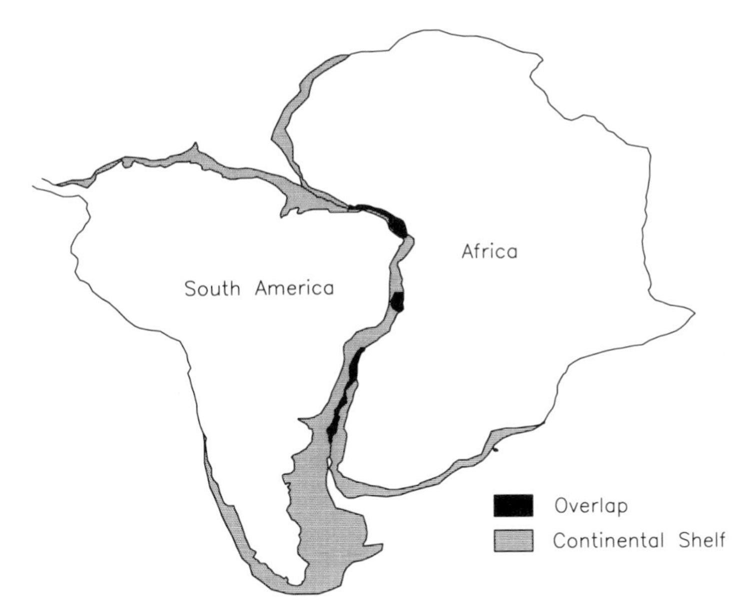

It all began because the coastlines of Africa and South America could be put together like the pieces of a jigsaw puzzle. Doing this gave rise to the idea that the two continents had once been part of the same landmass and had somehow been torn apart. The "fit" in this figure was done by Sir Edmund Bullard and associates with the aid of a computer. The areas of overlap are due to rivers having deposited large amounts of sediment off coastal areas, causing the continent(s) to increase in size.

The fixing and immutability of continents was the predominant view among earth scientists until 1915, when Don Quixote came riding over the geologic horizon to joust with the establishment. In that year a German meteorologist by the name of Alfred Wegener published his ideas on what he called "continental drift."

Wegener proposed that long ago all of the continents were joined together in a single landmass, which he named Pangaea (all lands). Over time this huge landmass broke up, and the various parts of it, our present continents, moved through the ages and across the earth to their present positions.

Wegener's writings caused only the breath of a stir until his work was translated into English in 1924. All of a sudden the proverbial fox was loose in the henhouse and the hens wanted no part of it, vigorously flapping and squawking to get rid of that fox and his ridiculous idea.

Alfred Wegener continued to write and talk about his "revolutionary" ideas until 1930 when, practicing his profession, he died in a blizzard in Greenland. (One anonymous earth scientist was heard to say that his weather predictions were about as good as his grasp of geology.) Like many people of vision and imagination, this nongeologist was scoffed at, or, at best, simply ignored. Wegener did not live to see the proof of his theory generate as great a revolution in earth sciences as the theory of relativity had caused in physics.

Alfred Wegener based his concept of drifting continents on four main pieces of evidence.[1]

First, starting with the continents of Africa and South America he proceeded to fit, in jigsawlike pattern, all of the continents back together again.

Second, not only could he "reasonably" fit the continents together but, like a real jigsaw puzzle, each piece, each continent, also contained part of a picture, a picture that made geologic sense when all the pieces were together. Wegener's picture was made up of rocks and mountains. He showed that the rocks that occur at or near the edges of the continents matched in type and age when he put the continents together. Also, mountain ranges that mysteriously ended at the edge of one continent could be shown to reappear suddenly on another continent. The Appalachians could be connected up the eastern United States, through Newfoundland to Greenland and down northern Europe. Same rocks, same age, same mountain range, with a large body of water called the North Atlantic now fitting in between.

Third, identical fossils, with similar ages in similar rocks, were being found on widely separated continents. How could a given life-form, be it animal or plant, develop simultaneously and identically on different continents now so widely separated?

To answer Wegener, it was proposed that the smaller animals swam, whereas the slightly larger ones were carried by logs or pieces of wood (these were the first boat creatures). For the really big animals there had to be great land bridges that connected the continents. One such land bridge was proposed between South America and Africa, a span that would have prompted an excursion longer than the incredible journey. Unfortunately, except for the known land link across the Bering Sea (Bering Straits), all of these land bridges had conveniently vanished beneath the ocean without a trace.

Fourth, there was the question and puzzle of climate. In general, the equatorial regions of the earth are the warmest areas, the polar regions are the coldest, and the areas in between have moderate temperatures. From about 1840 to Wegener's day, however, geologists had found evidence that seemed to contradict this and to do so in a big way. There was evidence of extensive glaciation in areas that are now subtropical (India, South Africa, South America). At the same time that these places were under the weight of thousands of tons of ice, areas like the United States, Europe, and Siberia had huge warm-water swamps choked with lush, green vegetation. This vegetation was eventually to rot and become buried to form the extensive coalfields now found across Europe and the United States.

In looking at the fossil and rock records it turns out that in many places the ancient climate of an area differed drastically from the present climate. Such climatic changes could not be explained by great global climatic shifts, for the rock record clearly shows that the same warm or cold trends did not occur simultaneously on all continents.

Alfred Wegener used this evidence to argue that the apparent change in temperature of a given area over time was due to continental creep, continents slowly moving from one latitude to another. Wegener went on to say that "this evidence is so compelling that by comparison all other criteria [of continental drift] must take a backseat."

In fact, it turns out that when the continents are "fitted" back together the glaciers that scoured ancient India, Southern Africa,

Position of the continents: (A) 200 million years ago when they were one; (B) today. Crosses indicate areas that were covered by glaciers when all the continents were joined together in a land called Pangaea.

eastern South America, and Australia all merge to form a great continental ice sheet. Well, Wegener had put forth his evidence. When the verdict was returned, the continents were found not guilty by reason of insanity—Wegener's insanity. "No mechanism," shouted the scientific community. Wegener had put forth no process to explain how continents could plow through ocean floors like great ice breakers through the arctic ice pack. Wegener had said that mountains would pile up at the continents' forward edges, and that the continents would leave a deep trail behind them, like crabs scuttling through wet sand.

Trying to explain how the continents actually moved across the surface of the earth, how they moved through the ocean floor, was Wegener's Waterloo. There was no topographic evidence of any dramatic disruption of the seafloor as would be expected by such a plowing or gouging process. There was no evidence that, and no mechanism known to explain how, solid continents could move over a solid, rigid earth.

Benjamin Franklin, who predated Wegener, had an inkling of all this. In 1770, 150 years before Wegener, between discovering electricity and writing *Poor Richard's Almanac,* Franklin proposed that the earth had a spherical core surrounded by lighter fluids. These fluids buoyed up the various continents, and "thus the surface of the globe would be a shell, capable of being broken and disordered by the violent movements of the fluids on which it rested."[2]

Arthur Holmes came even closer. In 1928, two years before Wegener's death, this Scottish geologist proposed that molten rock rose from deep within the earth. As the hot material neared the surface it cooled, causing it to slowly sink back into the depths. Holmes suggested that the rising and sinking movements (which he called convection currents) took place in a circular fashion, much like the tracks move on a D-9 bulldozer, and that this conveyor belt–like action was then responsible for pushing and pulling the continents over the earth.[3]

Detailed studies of the ocean floor in the 1960s and 70s showed that both Franklin and Holmes were not that far wrong. These studies were largely based on how seismic waves, or vibrations, generated by earthquakes or nuclear explosions travel through the earth. They clearly showed that our hard, rock solid earth is not as sure and strong as it appears.

LIVING ON PIECRUST

Imagine slicing the earth into pie-shaped pieces. Better yet, imagine a large piece of grandma's delicious apple pie. It's hot out of the oven and smells so good you just can't wait to cut into it. So

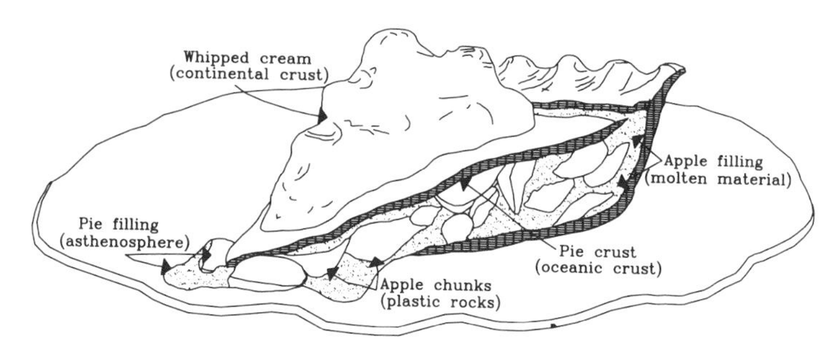

Grandma's apple pie and the parts of the earth. The piecrust and whipped cream represent the earth's lithosphere, which makes up only 0.1% of the planet's entire volume. The whipped cream represents the continents (continental crust), and the piecrust the ocean floors (oceanic crust). Grandma's hot, runny, but chunky apple filling represents the earth's asthenosphere, or "layer without strength."

you do—well before it has had time to properly set. As you put the piece on your plate the apple filling (with big chunks of hand-picked apples) slowly runs out onto the plate. As the filling flows, it actually drags the front and sides of the piecrust with it. If you add whipped cream to the top, then you have a pretty fair model of the very upper part of our planet.

The piecrust and whipped cream, together, are far more stiff then the hot apple filling. Together, piecrust and whipped cream represent the more rigid, solid outer layer of the earth. This is the layer geologists call the lithosphere (*litho* means rock or solid). The lithosphere makes up only 0.1% of the planet's entire volume, which makes it about as thick as the skin on an apple.

The whipped cream part of the lithosphere is light and airy and forms peaks that stand well above both piecrust and hot apple filling. The whipped cream can be compared to the continents and the material they are made of, called the continental crust by geologists. Like the whipped cream the continents are composed largely of light elements (such as sodium, potassium, aluminum, and silicon) and so they sit high above the piecrust, which can be used to represent ocean floors and the rigid material beneath them (called oceanic crust). Made up of heavy elements, such as iron and magnesium, the oceanic crust is much denser or heavier than our continental (whipped cream) crust.

The apple filling in our pie, which is equivalent to the layer below the lithosphere, is called the asthenosphere (without strength). This part of the earth extends down to an average depth of 300 miles (483 km), and lies entirely within the upper mantle (the zone between the earth's core and crustal material).

Like our hot pie filling this layer has no strength or rigidity. It can be compared to the apple filling in another way, for it consists of both hot, runny filling (melted rock, or magma) and hot apple chunks (soft, mushy, but not yet molten rock), with the apple chunks being far more abundant. With the high temperatures and pressures that occur at depths of 60 to 300 miles (97 to 483 km) this material, much like the apple filling, will flow or run sideways at a slow rate.

As it flows outward it drags the overlying rigid layer of the earth with it. Thus we can picture the light continents and the heavier oceanic material, being pulled and dragged by the mushy, slowly flowing rocks that make up the asthenosphere.

PIECES GALORE

If you've ever stood on the shore of an ice-covered lake in the early spring you would see that the ice covering the lake doesn't just melt and disappear as one huge mass. It is pulled, jostled, and broken into many smaller pieces. It is at the edges of these pieces, where they touch, that most of the grinding and crunching, the creaking and groaning takes place. Water may even spout out from between the pieces to shoot high into the sky like a geyser. The central portions of these ice pieces, however, remain relatively calm and undisturbed.

So it is with the earth's rigid outer layer—the continents and the ocean floors. They don't slide and creep over the earth as one huge monstrous mass. They are like the ice. They are pulled, jostled,

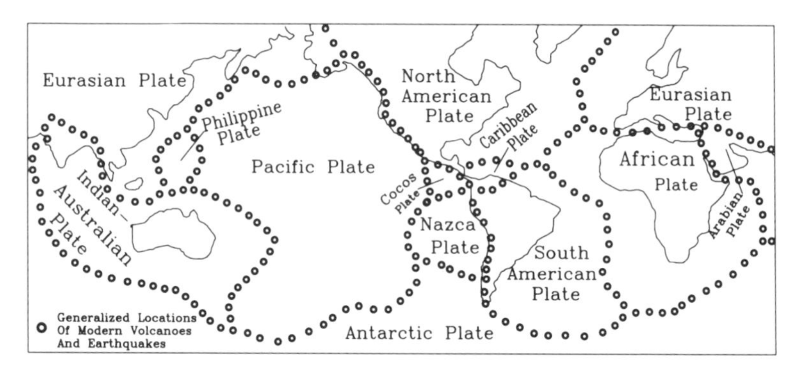

Generalized locations of modern volcanoes and earthquakes, which outline jigsaw-shaped pieces of the earth that geologists call plates. The six largest plates and a few of the many smaller ones are shown in this diagram.

and broken into many pieces. And it is where these pieces meet, where they touch, that most of the earth's action takes place. The central, thickest parts of the pieces are like a small Vermont town; they are relatively quiet and undisturbed.

What kind of action occurs here? If you take a world map and plot the locations of modern volcanoes and earthquakes on it, you will see that these are confined to elongated belts or linear chains (like the "ring of fire" that encircles the Pacific Ocean). By running your finger around these belts you can see that they outline oddly shaped pieces of the earth's surface that look like they belong to a giant jigsaw puzzle. What you have outlined are the places where our rigid earth, the lithosphere, is cracked or broken. The action taking place along these breaks is provided, free of charge, by volcanoes and earthquakes.

The oddly shaped pieces you have outlined are of all different sizes and are called plates by earth scientists. The study of the jostling and creaking of these plates is called plate tectonics. Tekton, the carpenter in Homer's *Iliad*, was the constructor, so tectonic has come to mean construction. Therefore, plate tectonics is the study of how these pieces of the earth are put together. So far six large plates and many smaller ones have been identified. The largest, called the Pacific plate, includes the southern part of California and the Baja peninsula. The rest of the North American continent forms part of the North American plate (what else?). These plates, or broken pieces, are equivalent to our piecrust (the heavier oceanic material), and they carry the lighter continents (the whipped cream) along with them.

MAGNETIC MYSTERIES

If South America (which forms part of the South American plate) and Africa (which forms part of the African plate) were once joined, then there should be evidence of their parting somewhere between the two continents. Continents breaking up should leave great scars and long, deep cracks to mark the site of separation.

If we look again at a world map, but now put on the topography of the ocean floors (as determined by echo sounding, seismic studies, and camera coverage), we see that the floor of the Atlantic Ocean has a distinct ridge that runs north-south—a great scar occurring about halfway between South America and Africa. This is the Mid-Atlantic Ridge, which, in its entirety, is more than 7,000 miles long (5,865 km). It is a great submarine mountain range which, like one seam of a baseball, winds its way up through the South and North Atlantic Oceans. Iceland, and its active volcanoes, represents one of the few places where the ridge rises above sea level. Similar ridges (called rises where they exhibit less rugged topography) have now been located on the floors of all the oceans and are, themselves, places of great volcanic activity and earthquakes. Are these ridges then the linear scars, the great cracks that mark places where piecrust is being dragged apart?

The material that composes these ridges, as well as most of the ocean floor (our piecrust), is dark, heavy, and made up dominantly of a rock called basalt. Basalt is a fine-grained volcanic rock that flows from volcanoes like those found on the Hawaiian Islands or Iceland. Basalt is made up of minerals that are rich in iron and magnesium. Certainly not as tasty as grandma's flaky piecrust, but when basalt comes out of the oven (erupts onto the ocean floor), it's a lot hotter. It's so hot that the iron- and magnesium-rich minerals are in a molten state called lava, or magma. As this hot material cools, individual iron- and magnesium-rich crystals start to form. This is similar to the formation of long ice crystals in a puddle of freezing water. Cooling and crystal formation turn the iron-rich minerals into tiny compasses. Just as a person turns until his hand-held compass points toward the magnetic north pole, so the iron in the minerals aligns itself parallel to the lines of force of the earth's magnetic field—the magnetic north pole.

Now, for the purpose of a proper explanation, it's necessary to take two steps backward. Step one: During World War II the Allies, trying to stop the constant attacks on supply convoys by German U-boats, developed what was called a magnetic airborne detector (MAD—like the war). This instrument, when lowered into the water from a ship or aircraft, could detect and locate the steel (iron-rich) hulls of submarines. After the war, in the 1950s, scientists adapted this instrument so they could tow it behind survey ships, enabling them to study the magnetic properties of iron-rich rocks on the seafloor.

And step two: Even before World War II, back in the 1920s, geologists and physicists knew that the earth's magnetic field flip-flopped back and forth, with the north and south magnetic poles switching places.[4] Exactly why this happens is not fully understood. But what it meant to the little compasses—the iron in the iron-rich minerals of basaltic rocks—was that they would point north when the earth's magnetic field is in its current position (called normal magnetism), and south when the poles are reversed (called reversed magnetism).

Why all this fancy footwork? The MAD surveys drove the scientists conducting them crazy and caused Wegenerian shock waves through the scientific establishment. The ocean floor was found to be composed of magnetically alternating bands (just like stripes on a zebra) of these iron-rich basalts. Within one band the iron in the minerals was aligned to point toward the current north

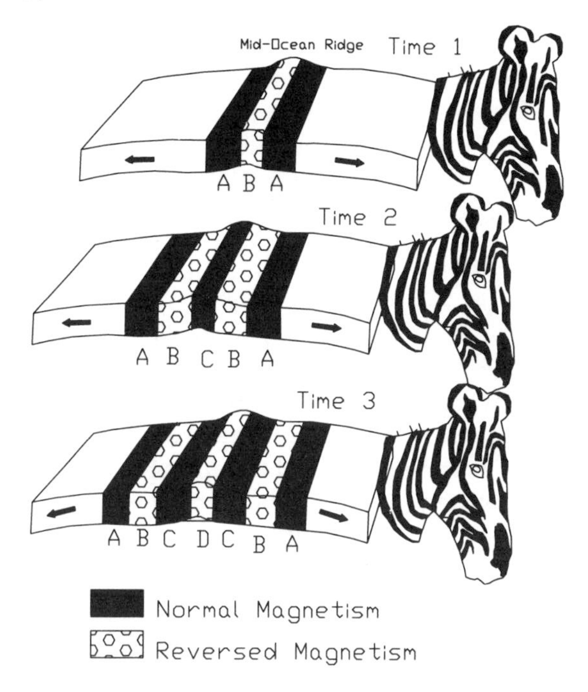

Zebra stripes across the ocean floors help illustrate the theory of seafloor spreading and moving continents. Lava extruded onto the seafloor at Time 1 shows reversed magnetism, whereas lava extruded at Time 2 shows normal magnetism. These flows are split apart at oceanic ridges and move laterally away to make the striped pattern observed across the ocean floors. This is a never-ending process, as shown in Time 3.

pole (normal magnetism), but in the very next band it was aligned to point toward the current south pole (reversed magnetism). And so it went across the seafloor, north to south and back again. Almost as amazing was the fact that these zebra stripes had arranged themselves symmetrically around the ocean ridges.

What had happened was that hot lava, upon cooling to hard, basaltic rock, had frozen into it a permanent record of the reversals of the earth's magnetic field. If we think about it, this is only possible if these basalts formed on the seafloor at different times or ages. These zebra stripes across our oceans gave two Cambridge geologists, Frederick Vine and Drummond Matthews, the evidence they needed to prove the ideas of Princeton geologist Harry Hess.

In 1960 Hess proposed the possibility that not only did continents creep, but the seafloors crept right along with them, that ocean floors actually split or cracked along places like the Mid-Atlantic Ridge, and then moved slowly away from these deep scars. These scars or cracks then became the places where hot lava rose up from the asthenosphere (pie filling) to erupt onto the seafloor. This lava then cooled to form new oceanic crust which, in its turn, was split apart to make way for a new batch of lava, the process repeating over and over, eon after eon.

Hess went on to propose that this creeping, spreading seafloor (our piecrust) eventually self-destructed by sinking back into the pie filling (asthenosphere), and, as it did so, it formed the deep oceanic trenches we see around the world (such as the one off the west coast of South America and the one along the Aleutian Islands). Unfortunately, Hess couldn't prove any of this, especially to the established scientific community that still scoffed at Wegener's ideas. So, to avoid the same fate, he quietly presented his ideas as nothing more than "geopoetry."

With the help of Vine and Matthews, Wegener's vision of drifting continents was introduced to Hess's geopoetry of seafloor spreading, and suddenly plate tectonics was a reality.

RECYCLING—EARTH STYLE

If the idea of seafloor spreading is to hold water, then the basaltic rocks found on the ocean floors should increase in age as they get farther and farther from where they were born—the mid ocean ridges. In fact, there should be a continuous age change across the ocean floors as we move from ridge to continent, or ridge to oceanic trench.

This idea was being proposed when the good ship *Glomar Challenger* set sail in 1968. During the next two years this oceangoing research vessel sailed between Africa and South America, systematically sampling the sedimentary material that covers the ocean floor and the basaltic rocks beneath this cover. Fossils found in the sedimentary material, and age dating of the basaltic rocks, showed not only that the rocks got older as distance from a mid ocean ridge increased, but also that the ages are distributed symmetrically on each side of a ridge: The basalt had been continually split and moved apart. This was indeed powerful confirmation of the ideas of continental drift and seafloor spreading. The continents did not have to bulldoze their way through solid ocean floors; the seafloors moved too and carried the continents along with them.

The earth is old and wrinkled, but some parts are much more wrinkled than others. The oldest rocks found on the ocean floors formed about 200 million years ago, just before the dinosaurs took over the planet. But these rocks are mere toddlers compared to the more than 3.5-billion-year-old rocks that form the heart and soul of most of the continents. More than 3.5 billion years of whipped cream but only 200 million of piecrust—it hardly seems fair. As it turns out, it isn't. The earth discriminates, and over the eons it has worked hard to preserve continental crust and rub out ocean floors.

This observation led to the final confirmation of continental creep and ocean floor spreading. If the earth is not growing bigger (no evidence for middle-age spread here), and if oceanic crust is continually being manufactured at midocean ridges, then the crust has to be consumed somewhere else. If it isn't, then it is difficult to explain the fact that no rocks older than 200 million years exist at the bottom of Jacques Cousteau's world. But where is oceanic crust destroyed and with what?

The answer arrived via a great mushroom cloud. To stop above-ground testing of nuclear weapons several countries (including the United States and former USSR) signed a treaty in 1963 limiting nuclear tests to the underworld. To ensure compliance 125 seismographic stations were installed at various sites around the world. Seismographs are instruments that measure the intensity and determine the location of seismic waves generated not only by nuclear explosions; but also by earthquakes. After 1963 these instruments provided a great quantity of data that showed a pattern of earthquake centers and intensities confirming that oceanic crust was indeed being gobbled up beneath the oceanic trenches.

The deep oceanic trenches marked the places where one jigsaw piece (a plate) was being bent or deflected down below another piece (geologists call this bending process subduction). The downgoing piece is eventually dragged to a depth where the temperature of the asthenosphere (pie filling) is high enough to melt the basaltic rocks. In this way basalt is slowly turned back into that from whence it came—hot magma, or lava. Old basalt into young lava, which is then recycled back through the crust.

This process has gone on over and over through the fog of geologic time, oceanic crust continually drinking from the earth's fountain of youth to be ever young and energetic. But what about wrinkled, arthritic continental crust, sitting there getting older and older, constantly worrying about putting on weight. Why is continental crust (whipped cream) older than oceanic crust (piecrust)? Remember, continental crust is made out of frothy, light and airy elements, so it acts like a cork on water and just goes bob-bob-bobbing along on the pie filling. It follows wherever the plates take it. It is just too darn light to drag or pull down into the asthenosphere. Every time the earth tries to rub out continental crust it pops back up as quick and easily as a jack-in-the-box. So continental crust gets older and older and has to contend with an expanding waistline. Over the eons the continents have gained ground, adding to their leading edges small, leftover scraps of oceanic crust or continental bits they happen to bump into. In fact, it has been estimated that the North American continent has added some 300 miles (484 km) to its waistline (all west of the Rockies) in the past 100 million years.

HOW THE PIECES MEET

The ads in the personal section of the *Earth Daily News* were startling but not unusual:

1. MCC (Mature Continental Crust), light and trim, a non-rifter, enjoys camping and fresh air, seeks younger, faster-moving BOC (Broken Oceanic Crust).
2. YOC (Young Oceanic Crust), fast paced and free, likes sailing and seafood, rifting no problem, a smoker who enjoys mineral water, seeks YOC with similar interests.

As you've suspected, the jigsaw pieces don't really meet like this, but the types of crusts that collide go a long way in determining what the crusts' ultimate fate will be—fatal attraction or love at first collision.

Breaking Up Is Hard to Do

It's tough being tugged and pulled apart, like a piece of licorice by two determined 10-year-olds. Slowly the licorice is stretched like a rubber band and, as it stretches, it thins until—snap!—both kids end up with a bit of thin, torn licorice.

The stretching, thinning, and breaking up of the earth's rigid lithosphere (piecrust and whipped cream) is similar. This breakup can take place entirely within oceanic crust (such as at the midocean ridges) or entirely within continental crust (as Africa and South America broke apart). Earth's history tells us that the breakup of oceanic crust is far more common, possibly because there is much more of it, and it is thinner and therefore easier to stretch and break.

In a general way we have already seen what happens when ocean floors move apart along midocean ridges. But what happens when two parts of a single continent find themselves incompatible?

The breakup is usually caused by a third party (doesn't that sound familiar?): hot basaltic magma rising up beneath continental crust. As the magma moves upward it pushes and shoves at the overlying continental crust, causing the crust to arch or bow like bread rising beside a warm stove. Arching causes numerous long, deep cracks to form in the crust. Then, as the hot magma moves upward and starts to cool (starting the tracklike motion of an Arthur Holmes' conveyor belt), the arched, broken crust is dragged and pulled apart (geologists call this rifting). This pulling apart causes a depression or valley to form (a rift valley), the bottom of which is filled by the cooled and crystallized magma that started the whole thing in the first place. What we have done is to create new ocean floor in the middle of a continent, and, if the process goes to completion, the valley (rift) will go on to become a new ocean. This dragging apart continues as long as magma rises beneath the valley and the conveyor belt moves on. This is how South America and Africa drifted away from each other 200 million years ago and how East Africa is being dragged away from the rest of the African continent today.

An ancient example of such a continental breakup can be found in the middle of North America. About 1 billion years ago the west got sick and tired of the east and tried to separate along a crack (rift) that extended from what is now Lake Superior down through Minnesota into Iowa and Kansas. Hot magma rose along this crack and spilled out onto the land for millions of years to form a great pile of basaltic rock. However, when the continent had been dragged apart about 30 or 40 miles (48 to 64 km) the process suddenly stopped, and what could have been never was—an ocean never formed, and Duluth, Minnesota, was not to be the French Riviera of the north. Today you can walk on the remnants of this great lava field. The basaltic rocks are exposed along the north shore of Lake Superior and the upper peninsula of Michigan and look little different from the basalt that today flows out of Hawaiian and Icelandic volcanoes.

This old scar, the so-called Midcontinent Rift, may be out of sight, but it's certainly not out of mind or hearing. The Midwest occasionally hears the grumble of a mild earthquake. This is the dynamic earth telling us about our past and letting us in on a little secret: Just because the continent was broken then sealed, doesn't mean it will stick together forever.

Convergent Plate Boundaries—Fatal Attraction

Where jigsaw piece meets jigsaw piece, the extent and form of the relationship depends on who is meeting whom and how they are introduced.

Continental crust (whipped cream) is frothy; it is light and airy, so it "floats" on the asthenosphere (our pie filling) like whipped cream on coffee. Oceanic crust (piecrust), like brandy added to the coffee, is heavy and wants nothing more than to sink into the asthenosphere (or pie filling). Which of these crusts is at the plate edges determines the fate of each meeting piece.

Like attracts like, so oceanic crust likes to meet oceanic crust. Unfortunately, this is also a fatal attraction, for one of the pieces ends up being bent or pushed down beneath the other. The "losing piece" is helplessly caught on an escalator that will carry it down deep into the asthenosphere. As it travels deeper into the "oven of the earth" the temperature increases and the basalt undergoes meltdown. The melted basalt, now hot magma, is then faced with two choices: It can rise back up through the overlying crustal piece to form long chains of island volcanoes (called island arcs, such as the Aleutians or the Marianas), or it can take a ride on Arthur Holmes' conveyor belt. Here it bides its time, waiting like the Phoenix, so it can again rise up to seal another great crack, starting the whole process over.

When oceanic crust (piecrust) meets continental crust (whipped cream) it's no contest. One-two-three and the oceanic crust is out. Being heavy, it simply slides below the continental crust and is quickly dragged down to be consumed in the earth's fiery cauldron. This melted material rises back to the surface as volcanoes located on the overlying continental crust. These form great chains of volcanic mountains such as the Cascades and the Andes.

Continents in collision at a convergent plate boundary. Like the meeting of two Sumo wrestlers, continental collision stands the crust straight up, slowly giving rise to high mountains. As the crust is forced upward it bends, buckles, and twists.

Continental Collision—When Sumo Wrestlers Meet

The two Sumo wrestlers enter the ring. Crouching, they watch each other like strange dogs meeting in an alley. Slowly they circle, grunting and flexing, beads of sweat glistening on their broad, oiled shoulders. In a sudden explosion of muscle they charge forward; flesh collides with flesh, and the earth shakes. The impact stands both wrestlers straight up. Straining and pushing they jockey for leverage and position—bulk against bulk—huge and proud.

India and Asia met in this manner. Buckling, bending, and twisting the crust, the collision stood the two continents straight up and the Himalayas slowly rose. Continent colliding with continent forms mountains of buckled, bent, and twisted rock; the Appalachians are mere remnants of such a collision, whereas the Alps, still tall, speak proudly of the collision between Africa and Europe. Mark Twain may have recognized something of this when he wrote, "Switzerland would have made a pretty impressive country if it was rolled out flat"[5] (which it once was).

Plate Hotspots—Acne of the Earth

Volcanoes from nowhere—pimples on plates far removed from where the action is (plate edges, or boundaries). For some mystic reason (which the earth knows but we don't) a plume of hot magma rises from the asthenosphere like an elevator going straight to the thirty-fifth floor. This elevator, however, is not going to stop. It's going to smash into, and then right through, the building's roof. The rising plume melts its way through the oceanic crust to form volcanoes in the middle of nowhere. The Hawaiian Islands represent such volcanos, which geologists call "hot spot" volcanoes. In fact, there is a great chain of extinct volcanoes that stretches from Hawaii to Midway, some 1,250 miles (2012 km), and the ages of the volcanoes increase as we move away from the big island of Hawaii. The hot spot, this rising plume, is stationary, and the oceanic crust just keeps moving over it eon after eon (sort of like having continuous heartburn), creating a great volcanic trail. As the plate moves an active volcano (such as Mauna Kea) becomes dormant and others rise to take its place (like Mauna Loa and Kilauea).

Personally, I like the Polynesian explanation of the phenomenon better. According to their legends the twin sisters Pele (the goddess of fire) and Namakaokahan (the goddess of water) wrestled with each other all across the Pacific Ocean. Pele would move from island to island making fire pits and Namakaokahan would chase her and put out the fires. Across the Pacific they struggled, leaving behind them the remnants of their battles—one extinct volcano after another—until they reached the island of Hawaii. Now, for the moment, Pele is winning.

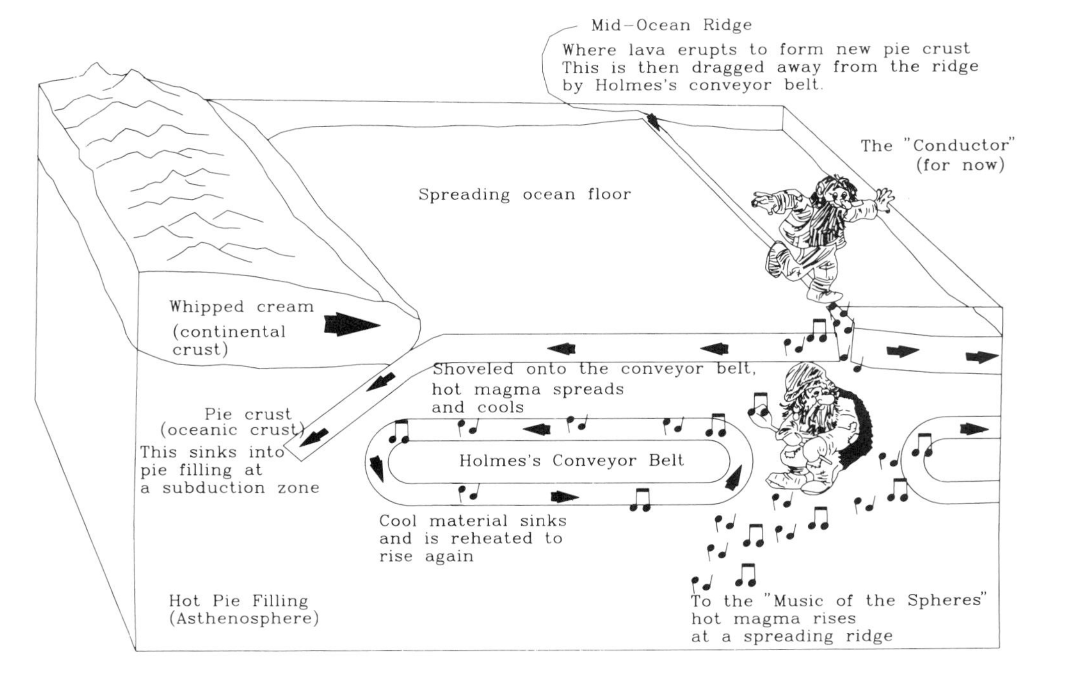

Possible driving force behind plate tectonics. Deep in the earth great forces toil to make the music of the spheres and to drive Arthur Holmes' conveyor belts.

Another known hot spot (and a future trouble spot) is the Yellowstone area in the western United States. A colossal volcanic eruption occurred here some 600,000 years ago forming a crater some 47 × 28 miles (76 × 45 kms) in diameter. Right now the continental crust under this area is 60 times hotter then the crust under the rest of the continent—hell's kitchen indeed.

ALL LANDS

Once the continents were together, then they drifted apart. Today they are like desert nomads; they have a great journey to make, with untold adventures ahead as they boldly go where no continent has gone before.

Two hundred million years ago Pangaea ("all lands") was a single continent. It has taken that long (the coming and going of the dinosaurs, the advance and retreat of great ice sheets, and the arrival of the mammals) for the continents to look the way they do and to get to where they are today. In that time India has had a head-on collision with Asia, the Red Sea has parted, Panama became the glue that stuck South and North America together, and Australia parted company with Antarctica. And that's not all.

By understanding earth processes we can look into the future and get a glimpse of where we will wander to. Imagine. Los Angeles and San Francisco pass each other like ships in the night, leaving Nevada as prime oceanfront property; South America, having had enough of North America, packs its bags and leaves, and poor Panama (canal and all) sinks into the ocean; East Africa drifts away from the rest of the African continent and Australia slams into China.

Deep in the earth, hundreds of miles from the madding crowd, great forces toil and struggle, making the music of the spheres. These forces drive Arthur Holmes' conveyor belts—hot magma rising, spreading laterally, and slowly cooling to sink back into the bowels of the earth to start the whole process over again. The rigid crust, our lithosphere—our towns, our houses, ourselves—dragged and pulled over the earth's surface from here to eternity.

Prelude to Chapter 2

Mount Pinatubo, St. Helens, Unzen; these were the images in my mind as I sat listening to the music. Deep, rumbling chords followed by sudden, roaring crescendos. Below this but rising in volume and violence were the explosions, continuous, one after the other, louder and louder. The music came from a great pipe organ; deep, loud, and continuous, it rattled off the stone walls and shook the windows. This was Pinatubo, St. Helens, Unzen, in all their violent glory.

CHAPTER 2

The Fire Within

Is it a God
Needing sacrifices,
A devil
Laughing at the souls of the damned,
A temptress
Charming and hungry,
A mountain of smoke and fire,
A place for dragons and wizards,
But look who comes,
Naked
To worship at its feet,
To plant and multiply,
To grow in its shadow
Until—

Darkness, swift and complete. The air fills with suffocating ash. The sounds—like "freight trains in thunder storms." Explosion after explosion, stirred and shaken by loud thumps sprinkled with whistles and the crack of hot rocks thrown through the darkness like bottles in a barroom brawl. Together they form an eerie beat, loud and vibrating—earth music fit for a great pipe organ.

Volcanoes and people. Together through the mist of myth, the richness of legend, and the dawn of understanding.

At the start it was to please and appease the gods. An Aztec priest, armed with a knife, the handle of gold and platinum, slaughters captives on the peak of a rumbling mountain. A young girl, colorfully dressed, bright flowers in her dark hair, is led by a village chief up a smoking mountain. At the top, trembling, she steps over the edge and for the moment the god is satisfied.

In other parts of the world it was to explain what was believed to be smoke and fire, and to give some understanding to the awful noise. The Roman god of fire, Vulcan, who was the blacksmith to all the other gods, worked his giant forges throughout Italy and the Mediterranean. The smoking mountains were the chimneys to his forges, and the rumbles and deep booming sounds were the god himself hammering away deep underground. In Indonesia a great snake, Hontobago, held up the whole world with its body. When Hontobago moved, the earth shook, fire flew from mountains, and even the gods were afraid.

Later on, volcanoes were used to lay blame at the devil's feet. Vulcan, in the Middle Ages, became a Christian demon, and his angry voice tried to drown out the moans and screams of torment of all the lost souls. Also, during this same period, it was largely believed that the Icelandic volcano Hekla was the gateway to hell, and the hissing chunks thrown out of it were the souls of the damned. In the sixteenth century many Europeans believed that Mt. Etna, on the island of Sicily, was the place of final confinement for the second wife of Henry the VIII—the poor, beheaded Anne Boleyn. There were those, however, who viewed volcanoes in a different way. Mark Twain was fond of saying that "the smell of sulfur is strong, but not unpleasant for a sinner."[1]

Always it was to try and explain the mountain. In the Pacific Northwest, so the Indian legend goes, two young chiefs, Wyeast and Klickitat, fell madly in love with the beautiful maiden Loo-Wit. The two chiefs begged Loo-Wit to choose between them—who would be her husband. But the maiden loved them equally and couldn't make up her mind. Soon the chiefs grew jealous of each other and fell to fighting. Their tribes also fought and many warriors were killed. The Great Spirit became so angry over the fighting and the foolishness that he turned the two chiefs into mountains: Wyeast into Mt. Hood and Klickitat into Mt. Adams. Out of compassion he allowed Loo-Wit to keep her great beauty. The Great Spirit turned her into the "Fuji-yama of the west"—Mount St. Helens.[2] She was indeed beautiful until the spring of 1980 when she threw an awful fit. She blew her top and her flowing tresses will never be the same again.

Finally, it was to try and understand how and why a volcano works. Plato believed there was a great river of fire inside the earth and hot air escaping from it caused earthquakes and volcanic eruptions. Benjamin Franklin thought the earth's crust was a thin, solid shell floating on lighter fluid. Movement of this fluid cracked the shell, causing earthquakes and volcanoes. Most recently it is moving plates. Where Alfred Wegener's creeping continents and Harry Hess' spreading oceans split apart or bump together, volcanoes may rule the land.

Why are we humans drawn to volcanoes like kids to a haunted house, like mice to a cheese-baited trap?

We all know volcanoes can be beautiful. Take Mount Fuji (it's been said there is no volcano on earth like it), or Kilimanjaro or Rainer or a hundred more—the incredible scenery, the lurking power. And, of course, nothing draws us closer to something than good old human curiosity. Just look at all the volcanic national parks there are: Lassen, Crater Lake, Yellowstone, Rainier, Hawaii, and Craters of the Moon, to name but a few. Millions of people each year enjoy volcanoes; they come for the beauty, the ruggedness, the allure—they come to see and to know.

Then there is the dirt. The soil around volcanoes is some of the richest, most fertile in the world, coming from the breakdown of ash and lava, natural fertilizers rich in potassium, phosphorous, magnesium, and sulfur. Field upon field, crop after crop grows around the mountain and up its lush sides—rice, bananas, coffee, pineapples, grapes, sugar cane, and much more. In Java the vol-

A volcano is defined as the opening or place from which molten rock and/or gas issue onto the earth's surface. The photo is of a fire fountain eruption of the Hawaiian volcano Kilauea. (Photo by H. Gibson.)

canic soil is so fertile that with enough rainfall the people there can harvest three crops a year.

Volcanoes are also natural energizers, and they just keep on going and going and going. Volcanic heat produces steam and hot water to drive turbines, to make electricity—to produce energy. New Zealand gets about 10% of its electrical power from such heated water (called geothermal energy). Volcanic energy is also harvested in Italy, Mexico, Iceland, Japan, the Philippines, and in the Geysers area of northern California.

Then there is the land. Volcanoes create not only fertile soil but also new land. All of the oceanic islands from Iceland and Hawaii to the Philippines and New Guinea are volcanic in origin. Much of the heart of our creeping continents, our ancient crust (older than a billion years) is of volcanic origin. The northern parts of Minnesota, Michigan, and Wisconsin, large chunks of Ontario, Manitoba,

Steam and gas coming out of the crater of the White Island volcano off the coast of New Zealand.

and Quebec, pieces of South Africa, India, and Australia are made up of very old, mostly worn away volcanoes. Volcanoes increase the earth's land surface, making a bit more room for all of us.

Last, but not least, volcanoes are places where many economically important elements of the earth are transported and concentrated to provide a farmers' market of mineral resources. Minerals that allow us to design and make computers and cars and planes and microwaves and just about everything else. Resources that allow us to explore outer and inner space, and let us live many steps removed from our cave-dwelling ancestors.

People around volcanoes. People enjoying volcanoes. People multiplying, thriving, growing until, house next to house, suburb crammed against suburb, cars bumper to bumper, the earth rumbles, the trap is sprung, and "thar she blows."

"History of any one part of the earth consists of long periods of boredom and short periods of terror."[3]

A volcano is also defined as the hill or mountain built up by the accumulation of cooled magma. Such a volcano is Mt. Ngauruhoe in New Zealand.

THIS GOD THAT STANDS SO HIGH

The word *volcano*, from the Roman god of fire, Vulcan, has two meanings. A volcano is the opening or place from which molten rock (magma) or gas issues from the earth's interior onto the surface. A volcano is also the hill or mountain built up around such an opening by the accumulation of cooled magma, which has been either poured or thrown out of the opening. So in a sense volcanoes make volcanoes, or is it the other way around?

The opening through which molten rock or gas reaches the surface can be either a circular, pipelike channelway (called a vent) or a long, skinny, snakelike crack (called a fissure or a rift). Both of these openings connect to a body of molten rock at a depth (a magma chamber).

The difference in shape between the two types of openings leads to differences in the final shape of the hill or mountain formed. Material that is poured or thrown out of a small vent opening piles up relatively close to that opening, and constructs a steep-sided, tall volcanic hill or mountain like St. Helens, Adams, Pinatubo, Fujiyama, and Kilimanjaro. These types of volcanoes are most common where jigsaw piece meets jigsaw piece (along plate edges).

Material that flows or is thrown out of a long crack will be much more widely dispersed and will form plateaus or plains, such as the Snake River Plain or the Columbia River Plateau, or low-angled hills and mountains, such as Mauna Loa and Kilauea in Hawaii. Such plains and mountains are common along midocean ridges and above oceanic and continental hot spots.

The sizes of vents vary from those no bigger than a sewer cover to ones that are larger than the Mall of America (parking lot included). Fissures can be as thin as a pencil or as wide as a football field, though most are about the width of a city sidewalk. Though relatively narrow these cracks can be quite long, even longer than a Minnesota winter. Some of the longer fissures observed in action are the Great Crack on Kilauea, which is 8 miles long (13 km), the Eldgji ("old fissure") in Iceland, with a length of 18 miles (29 km), and the Icelandic fissure that fed the great Laki lava flow of 1783, at 15 miles long (24 km). Some of the fissures that led to the formation of the Columbia River Basalts, the Keweenawan Lavas of Minnesota and Michigan, and the Deccan Traps of India may have been even longer. Finally comes the longest of the long, the snake of all snakes—the midocean ridges where oceanic crust is steadily being pulled apart. Possibly the Indonesians weren't so far off with their idea of the great snake Hontobago holding the world on its back.

NOT OF FLESH AND BLOOD

Molten rock that pours out of a volcano is called lava, and like maple syrup running off a stack of pancakes, it flows away from the opening. This lava comes from Vulcan's forge, or better yet, we can think of it as coming from grandma's kitchen. Only now grandma's into making bread. Measuring and kneading, letting it rise, then baking it so it can pour out of her volcanic oven, hot and fresh. In grandma's kitchen there are two kinds of bread: brown bread, which is made up of the same material as the oceanic crust (our piecrust), and white bread, which is the same stuff as the continental crust (our whipped cream).

The brown bread is dark and rich because it is made up of iron- and magnesium-rich minerals, called *mafic* by geologists (magnesium + ferric [iron] + ic). The white bread is composed of airy and light minerals (in both weight and color) rich in sodium, potassium, and silica, and has been given the name *felsic* (feldspar [sodium- and potassium-bearing minerals] + silica + c—the grade the geologist got for making this up).

Having been health conscious for more than 4 billion years, grandma bakes mostly brown bread. In fact, about 90% of the time brown (or mafic) lava comes out of her ovens, leaving a mere 10% white (felsic) for the kids to spread peanut butter on. And grandma's oven is a lot hotter than yours or mine. Her brown bread is poured out at temperatures of 1,858° F (1,000° C) to more than 2,308° F (1,250° C), whereas her white bread erupts at temperatures of 1,318 to 1,678° F (700 to 900° C).

Grandma's bread is so hot when it come out of the oven that it flows; lava is said to flow away from volcanoes. But using the words *flow* and *lava* in the same sentence is very misleading; lava is not slippery like water or oil or soda pop. In fact, lava, white or brown, is very, very sticky stuff, just like bread dough. You know how bread dough sticks to your fingers or the sides of a bowl? If you plop it down on a tabletop, it just lies there not doing much of anything—like a dog on a hot summer day. Same with lava. It's lazy, and this laziness, this resistance to flow, is called viscosity.

Fluids that flow or pour, like water, are said to have a low viscosity, whereas those that creep along, like molasses, are said to have a high viscosity. Lava, brown or white, has a huge viscosity, say 100,000 to 200,000 times that of water.

That's about the same as cement that's just about set, or teenagers showing a great resistance to doing homework. Yet you read about "rivers of lava" or "streams of lava" and hear phrases like "lava that flows like water." Well, it doesn't and it can't. But lava is a great impersonator and can give the impression of having the viscosity of motor oil. It's able to do this because of its high temperature and its great weight or density. Lava is 2.5 to 3 times heavier than water so, like Jabba the Hutt in the *Star Wars* movie *Return of the Jedi,* it can move or flow under its own weight. Moreover, the steeper the slope it flows down, the faster it can travel. Lava has been clocked at speeds in excess of 50 miles per hour (80 kmph), but fortunately for us most lava flows are the tortoises of the volcano world. They are slow and sluggish (average speed may be 4 to 5 miles per hour [6–8 kmph]), but like the fabled tortoise they are determined and persistent. They will roll over and bury anything and everything that gets in their way. So the great hazard of flowing lava is the damage it does to property and cultivated land.

The distance lava can flow is determined both by the steepness of the ground it's going over, and by how fast grandma can shovel it out of the kitchen. Flows of mafic lava in Hawaii can extend for more than 35 miles (56 km); some flows in Iceland are more than 80 miles long (129 km), and lavas in the Columbia River region extend more than 100 miles (161 km) from their openings. However, most flows are in the 5- to 25-mile (8 to 40 km) range.

Felsic lava is much stickier than mafic lava so, like visiting grandchildren, it doesn't move far from grandma's kitchen. Felsic lavas are usually ½ mile (0.8 km) to 5–8 miles long (8–13 km), and many actually pile up right over the opening to form bowl-shaped features called lava domes. Lava domes range in height from human size (a few feet tall) to ones that dwarf the World Trade Center. Most, however, are like sumo wrestlers—broader than they are tall. It is also not uncommon for these sticky lava domes to line

Dome-shaped accumulation of sticky lava. Shown here is the Merapi lava dome in Indonesia. The dome is covered by "crumble breccia," loose rock fragments formed by fracturing and "crumbling" of the top of the dome as it expands and/or grows outward.

up along a crack or fissure, like birds on a telephone line, to form a row of several domes. Chaos Crags at Lassen and Mono Craters at Mono Lake in California are examples of this.

The major volcanic hazard presented by lava domes is having them collapse and slide down the side of a volcano right into your Sunday afternoon barbecue. If a growing dome is jiggled by an earthquake, soaked with heavy rain, or becomes too big to support its own weight, the hot dome may come crashing down the side of a volcano at speeds in excess of 100 miles per hour (161 kmph), and it's not stopping for any traffic lights either. Avalanches like this are common and have been reported from domes associated with many of the volcanoes that form the so-called ring of fire around the Pacific Ocean.

Hot lava that is thrown out of a volcano by an explosive eruption is called pyroclastic (fire broken). It explodes out instead of flowing because grandma has gone to visit the grandkids and, in her hurry, she put way to much yeast in the bread dough. The extra yeast causes too many gas bubbles to form. As the dough slowly rises the bubbles become more and more active and agitated; the gas acts like carbon dioxide in a can of warm, well-shaken pop.

Inside a volcano, as this gas-rich dough (lava) slowly rises toward the surface, the active and agitated gas bubbles grow bigger and bigger. Their rapid growth, caused mostly by decreasing pressure, results in a violent frothing that rips and tears the lava apart. Take the cap off such a gas-rich magma and, like Loo-Wit, it blows its top. Set free, the shredded lava races up through the volcano and out the top, forming a towering column. This "eruption column" can extend for more than 10 miles (16 km) above the volcano.

The bits of blown-out lava are not unlike overrisen bread dough, for they are full of holes and very light. Called pumice, these bits come in a variety of sizes from crouton and doughnuthole to dinner roll and bread loaf, though most are around the size of crumbs from a crushed bread stick (this size material is called volcanic dust). Pumice, which is often light enough to float on water, is the main material of most explosive eruptions.

The bits and pieces of blown-out pumice can do one of two things. First, they can fall or rain back to the ground. The bigger bits, being heavier, land closer to the opening, whereas the lighter, smaller pieces are carried away by the prevailing wind to fall farther and farther from the volcano. Very tiny bits (dust size) can be blown right up into the stratosphere (10–15 miles [16–24 km]), where they can remain for several years. This "fallout" or "air fall" material is called volcanic ash and collects on the ground in even layers (or beds), like snow on a window sill.

The amount of ash generated by an explosive eruption can be incredibly large and deadly. The ash can cause total darkness at noon and can wreck havoc with machinery, telephone and electric wires, water supplies, and crops; it can emit poisonous gas, collect on roofs until they collapse, and start fires. In A.D. 79 the ash fall from the eruption of Mt. Vesuvius buried the town of Pompeii 10

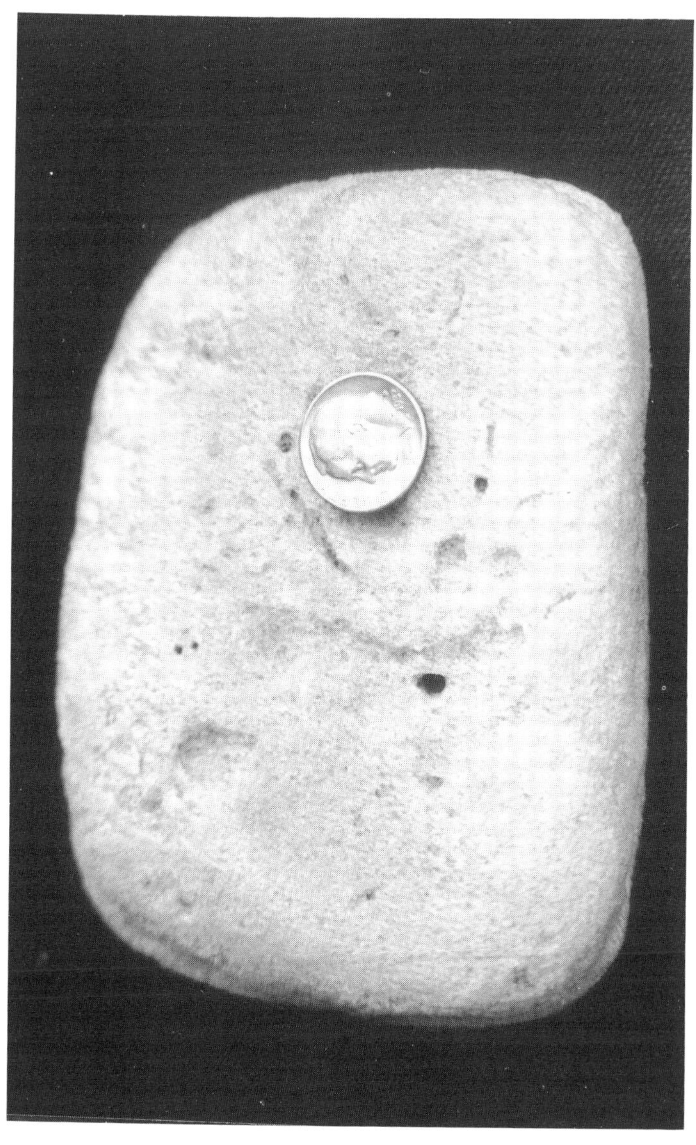

A breadroll-size piece of light, frothy pumice. (Photo by P. Morton)

feet deep (3.28 m), and at nearby Stabiae the ash filled the courtyards of houses. To date, 2,000 human remains have been excavated at Pompeii, all killed when the roofs of houses collapsed. The eruption of Mount St. Helens in 1980 deposited more than 600,000 tons of ash on Yakima, Washington, where snowplows had to be used to remove it from the streets. At Pinatubo, located on the island of Luzon in the Philippines, thousands of houses collapsed because of ash accumulation on the roofs, and over 300 people were crushed to death. Over 100,000 acres of good cropland was also buried under the ash during the 1992 eruption.

The second choice the airborne bits have is to try to stay together. At times there is so much pumice blasted into the air that it is much too heavy to be carried far up into the eruption column. So it falls back to the ground en masse, where it proceeds to speed away from the volcano like Han Solo's *Millennium Falcon* going into warp drive. This mixture of pumice and volcanic gas is said to "flow" (because it travels along the ground) and, like the devil, has been given an incredible number of different names. The most common are pyroclastic flow, ash flow tuff, and nuée ardente (French for "glowing cloud"). These fire-broken flows of pumice are hot, full of gas, and travel at high speeds. Faster than the Road Runner, faster than a speeding bullet, they charge over the ground at 100 to 250 miles per hour (161 to 400 kmph). During the 1980 eruption of Mount St. Helens, seismic stations 6 miles (10 km) from the eruption center quit sending signals 1.3 minutes after the eruption started. The cloud was traveling at 220 to 250 miles per hour (354 to 400 kmph). The 30,000 people killed by a pyroclastic flow at St. Pierre on the island of Martinique in 1902, had about 1½ minutes to escape the town. This flow, from Mt. Pelée, traveled 12 miles (19 km) at speeds greater than 90 miles per hour (145 kmph).

Not only are pyroclastic flows fast but they are also able to go a long way on a single tank of volcanic gas. They can travel anywhere from 5 to 150 miles (8–241 km) away from the volcano. This makes a pyroclastic flow an extreme volcanic hazard. Over the years such eruptions from volcanoes like Mt. Pelée, Leamington,

This photograph of the 1975 eruption of Mt. Ngauruhoe, New Zealand, shows an eruption column towering above the volcano and a flow of pyroclastic material moving down its side. This flowing material has fallen out of the eruption column. (Photo by M. Lambert)

A pyroclastic flow similar to the one that destroyed the town of St. Pierre, killing 30,000 people. This particular nuée ardente was from another eruption of Mt. Pelée, also in 1902. (Photo reproduced from the Library of Congress collection)

"Condominiums" dug out of soft, pumice-rich deposits left by the 1-million-year-old pyroclastic flows of the Bandelier Tuff. These particular dwellings, located in northern New Mexico, were carved out by the Anasazi Indians some 600 years ago.

Merapi, St. Helens, Unzen, Pinatubo, and Ilopango, to name but a few, have taken the lives of thousands of people.

On the good side, deposits left by pyroclastic flows have given shelter to hundreds of people over the years. Early Americans, such as the Anasazi people of New Mexico, Utah, and Arizona, not only built houses of masonry, but also dug many rooms, apartments, and condominium-like dwelliings out of the soft, pumice-rich material.

TURTLES, TEAPOTS, AND BATHTUBS

The hills or mountains built up around volcanic openings come in several different forms and shapes, but three are ex-

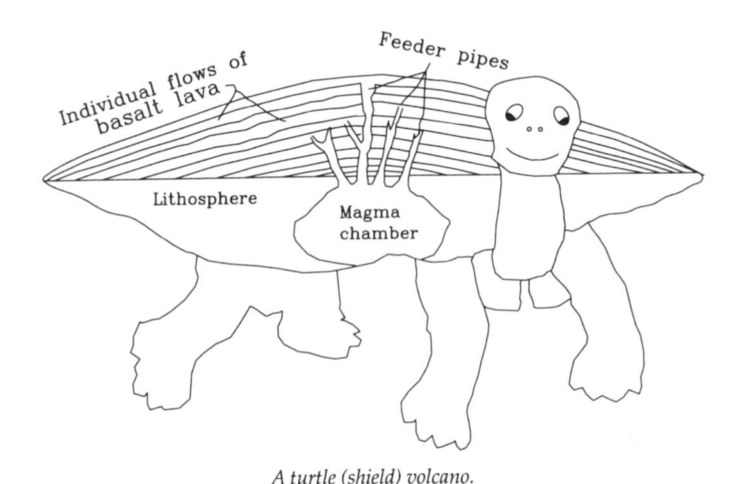

A turtle (shield) volcano.

tremely popular with planet earth. These are turtles, teapots, and bathtubs.

Turtles

Volcanoes shaped like turtle shells occur over oceanic hotspots and along spreading ocean ridges; Hawaii, the Galapagos Islands, and Iceland are all places where great turtles dwell.

> The burning lava soon overflows the brim,
> In streams of fire upon the sides expands,
> To desolate and spoil the blooming ground,
> A soil fertile with glowing rocks to fill.
> —C. S. Rafinesque

Turtle, or shield, volcanoes have gentle slopes of less than 10 degrees and are composed almost entirely of mafic lava (basalt) that has flowed out of cracklike openings. Even with such subdued profiles, they form some of the tallest mountains on earth. If we could see them in all their glory they would be awesome; however, we see only the parts that rise up out of the sea. Mauna Loa, on the island of Hawaii, has a basal diameter of 62 miles (100 km) and stands 13,685 feet (4,172 m) above sea level. But this is only half the story, for it rises another 16,400 feet (5,000 m) just to get to sea level. In all, the volcano is about 6.2 miles (10 km) high, making it taller than Mt. Everest. It is estimated that it took about 1 million years and thousands of individual lava flows to build this awesome beast. Other good-looking turtle volcanoes include Mauna Kea and Kilauea (home to the fire goddess Pele). If you want to see the granddaddy of all shield volcanoes you will have to become an astronaut. Olympus Mons, with a basal diameter greater then the length of the entire Hawaiian Island chain, lives and breathes smoke and fire on the planet Mars.

Mafic lavas that form turtle volcanoes are of two kinds. First is a low viscosity, smooth to ropey or shelly lava that Hawaiians call *pahoehoe* (puh-HOY-hoy, meaning "surface on which you can walk with bare feet"). As *pahoehoe* flows over the ground, like a river looking for a home, the upper part of the lava flow is air cooled and becomes stickier than the hot center. Once sticky, always sticky, and all this upper part wants to do is get out of the rat race, slow down, and enjoy the trip. Not so the hot center. Eager and ambitious, it wants to charge ahead and tries to drag the lazy upper part along with it. This dragging action, this pulling, twists the sticky top into long, curving, ropelike features (called ropey *pahoehoe*), or, if it is sticky enough, actually breaks it into slabs or shell-like pieces.

The other kind of mafic lava, is called *aa* (AH-ah), and that's exactly what you yell if you try to walk over its surface barefoot. The Hawaiians obviously tried this, for that's what *aa* means—"lava that cannot be walked on in bare feet." When you see it you know why. *Aa* begins life stickier than *pahoehoe* and therefore is sluggish and slow, and piles up at its front like waves coming in on a long, sandy beach. This leads to a surface that is rougher than #1 grit sandpaper, an angular, clinkery surface with tall, jagged lava spikes sticking up out of it. Because of this difference in "stickiness," *aa* crawls over the ground like a Sherman tank, whereas *pahoehoe* appears to flow smoothly down slope like a velvet ribbon of fire.

The sticky top of a pahoehoe flow from Kilauea, twisted into long, curving, ropelike features. (Photo by H. Gibson)

Both old smoothie and old clinkery work for the temperamental goddess Pele, so they can either flow calmly out of a fissure or charge out like bulls at a rodeo. The red-hot lava can spurt and gush hundreds to thousands of feet up into the air to form spectacular fountains of fire. At night they glow and pulsate, burn bright red or a cinnamon brown, and dance and flicker, moving to the rhythm and roar of the fire goddess, Pele. These fountains of fire feed glowing flows of lava that move, like coiled snakes, down the side of the mountain.

Both *aa* and *pahoehoe* have gas. However, these mafic lavas are not as sticky (viscous) as felsic lava and so the gas (mostly water vapor, carbon dioxide, and sulfur) slowly passes out of the lava, leaving behind holes that look a lot like the ones in Swiss cheese. If the gas is passed while the lava is still flowing, the holes will be angular and contorted like a pretzel; if it is passed after the lava has stopped moving, the holes will be circular, like a doughnut hole.

Round vesicles (gas cavities) in a sample of a pahoehoe lava from Kilauea. (Photo by P. Morton)

Called vesicles, the holes vary in size from pinheads to the rare watermelon-sized hole. Over the years, as the lava cools and turns to stone, groundwaters seeping through the rock precipitate out minerals that fill the gas holes. It is in this way that beautiful thunder eggs, geodes, and agates are made. The holes in the Keweenawan Lavas of Michigan are filled with native copper, while those on the north shore of Lake Superior are filled by banded

Vesicles (gas cavities) filled by the minerals calcite and thomsonite. This sample is from the North Shore Volcanics of the Lake Superior region of Minnesota. (Photo by P. Morton)

quartz (agates), as well as other strange but pretty minerals like thomsonite and analcime.

Small, cone-shaped features may be seen on the sides of big shield volcanoes. These are called cinder cones because they are made out of mafic pumice (frothy, blown, out bits of basalt) that has the shape and feel of cinder (similar to coal lumps after being burned and cooled). Cinder cones have steep sides, are less than 1,000 feet high (305 m), and are like deadly parasites, for they mark

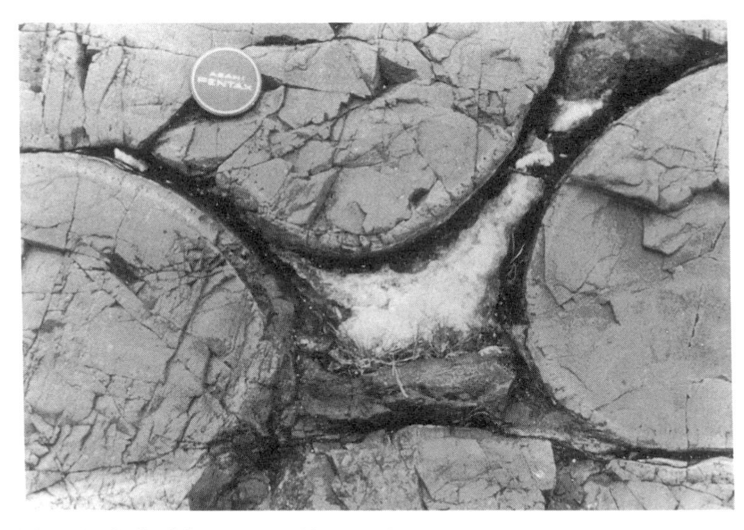

Submarine "pillow" lava, so named because of the shape the hot lava takes when cooled by seawater. More than 2.7 billion years old, this pillow lava can be found in Shebandowan, Ontario. (Photo by P. Morton)

the death of an old turtle. Their formation tells us that the lava has become sticky enough to be blown out; it can no longer flow. This means the heat source, the magma chamber, is either becoming ice cold and dying, or packing its bags to move next door.

This process is currently happening in the Hawaiian Islands. Volcanically active Kilauea, carried northwestward by the Pacific plate, is sliding slowly away from its heat source, the hot spot responsible for the formation of all the Hawaiian Islands. Over the next few thousand years volcanic eruptions at Kilauea will become mildly explosive, and cinder cones will form on the volcano like warts on a toad. Meanwhile, volcanic activity will increase southeast of Kilauea, and a new Hawaiian Island, already called Loihi, will rise out of the sea.

Lava that forms the underwater part of shield volcanoes looks and acts entirely differently than the above-water lava. When hot mafic lava meets cold seawater, the lava undergoes great change.

Few-million-year-old pillow lava found outside of Santa Barbara, California.

Cold water quickly cools the lava, which causes a rapid increase in the lava's stickiness. This makes the lava slow down and take on shapes that look like fat Albert's fingers—short, puffy, and bulbous. But, instead of ten fingers, there are hundreds, all intertwined like pieces of spaghetti. This lava is called pillow lava because the ends of the fingers are oval, and look like the pillow sitting on your bed. If the cooling is rapid enough, fat Albert's fingertips (the pillow ends) can snap right off, and tumble down the seafloor to collect in a great jumble that looks like the remains of a summer camp pillow fight.

You can actually stand on the submarine part of turtle volcanoes in such places as northern Minnesota, Ontario, Quebec, Manitoba, Michigan, and Wisconsin. Here, ancient shield volcanoes are partly preserved and now exposed at the surface. These vary in age from 1.1 to 2.7 billion years, but they still have the look and feel of such a one as Mauna Loa.

Water can do much more than just cool hot lava. It can put it under so much stress that the molten "rivers of fire" shatter into zillions of small, blocklike pieces. It's like taking a big rock and chucking it through the back window of your car. Starting from the impact point the crack expands, inch by inch, forming a webwork of loosely connected, tiny square pieces. Push on these and you have enough glass bits to clean up (while muttering under your breath) for a whole afternoon. This is exactly what happens to the lava, only sometimes the sea polishes up the blocky pieces and puts them ashore as black sand beaches.

Then there are times when the lava pours out of the fissure so fast, and is so hot, that the water doesn't get much chance to interact with it. Such lava behaves just like *pahoehoe* lava that is erupted at the surface. Down in Davy Jones' Locker, however, geologists call this lava "sheet lava" or "sheet flows" (I suppose they had to have something to cover all those pillows).

Teapots

Teapot volcanoes tend to steam and whistle and blow their tops a lot. This being the case, I thought it would be appropriate to introduce them by means of a children's poem called "The Ice Queen":

> A teapot volcano of some renown,
> Sat high above the little town,
>
> So majestic, so serene,
> The people called her the Ice Queen,
>
> They skied upon her slopes so steep,
> And grew their food around her fertile feet,
>
> They dug gold from her rocky side,
> And used it to build a city 10 miles wide,
>
> And over the years the people forgot,
> They forgot what they surely should not,

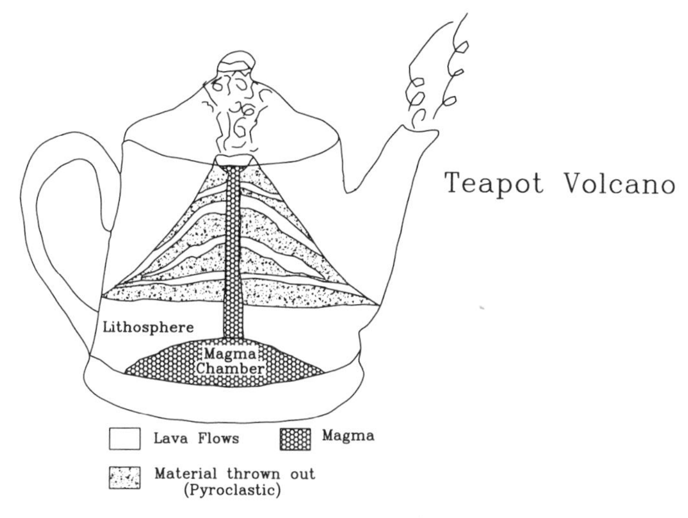

A teapot (composite) volcano.

They ceased to remember that when she was young,
The Ice Queen was wild, as wild as they come,

With gigantic explosions she blackened the sky,
And rained down ash from a cloud that went 30 miles high,

Hot gassy flows she sent over the land,
And buried it deep with pumice-like sand,

She bubbled and steamed from many a vent,
And poured out lava that flowed just like cement,

Oh, my, she would cry, the explosions are grand,
To rumble and roar and destroy all the land,

But quiet she's been for many an age,
So no one remembers when she was the rage,

For memories fade, people grow old and do die,
But not so the Ice Queen, she's far too sly,

A year to her is merely a blink,
A century or three only a wink,

And then her alarm, in the form of a quake,
Trembled and rattled and caused her to wake,

The mountain exploded in one great flash,
Four seconds later the city was ash,

And one million people who'd forgotten the Queen,
Were baked like lobsters in hot sulfur steam,

Oh my, she did sing, aren't the eruptions just grand,
To rumble and roar and destroy all the land,

When it was over, the politicians came around,
Was one chance in a billion, we'll rebuild our town,

And the people they ski down her slopes so steep,
And live in the new city they built at her feet,

And what spins round, will come round again,
The Ice Queen will surely commit the same sin,

And that's the end, as far as I know,
For the Ice Queen once more is all covered in snow.

Teapot volcanoes are made of lava that has both flowed and been blown out. For this reason teapot volcanoes are called composite volcanoes, or strato-volcanoes (*strato* means the volcano is layered, or stratified, like lasagna). These volcanoes form where jigsaw piece meets jigsaw piece, giving us great chains of fiery continental mountains, such as the Cascades in the western United States, and long lines of exploding islands like the Aleutians, the Azores, or the Canary Islands.

Most of the eruptions from teapots are out a central spout, and the lava is a lot stickier and cooler than that which builds turtle volcanoes. For these reasons teapot volcanoes form tall, steep-sided mountains (slopes of 15 to 35 degrees) that can be more or less symmetrical and very beautiful, attested to by the likes of Fuji, Kilimanjaro, Tongariro, and Hood. These volcanoes show a terrible tendency to do violence. They are among the world's most dangerous and have proven this again and again over the last few thousand years, with destructive eruptions from the likes of Krakatoa, Pinatubo, Pelée, Vesuvius, Augustino, Hekla, and Unzen. Today the world's tallest teapot is called Queen Maries Peak in Triston da Cunha (and just where is that, Carmen Sandiego?). It rises 3.1 miles (5 km) above its base.

Composite volcanoes are made up of short, sticky lava flows, fallout ash, and glowing avalanches or pyroclastic flows. These various volcanic products commonly form alternating layers, like the layers of cream, chocolate, and cherry filling in a Black Forest torte. However, in teapots, the relative proportion of lava flows to explosive material varies widely. This makes the eruptions difficult to figure out. They can be violent and disastrous (like Pinatubo), or they can be effusive and largely property destroying (like Etna).

Sometimes eruptions from these volcanoes fool the scientists who study them. Loo-Wit (Mount St. Helens) did this in 1980 when she erupted in a way that was totally unexpected. Her big eruption was triggered by a earthquake, which started a massive landslide. The landslide removed so much of the upper part of the mountain that it decompressed the magma chamber. It was just like taking the top off a warm, well-shaken bottle of pop. The decrease in pressure sent the magma crazy; it boiled and frothed all the way up the teaspout. The lava, however, didn't exit via the spout. It found an easier way. The shredded material came blasting out of the scar left by the landslide. This formed a lateral or horizontally directed pyroclastic flow that sped outward for 18 miles. In a matter of tens of seconds more than 60 people were dead.

Like the pharaohs of Egypt this eruption left a great monument to itself. Where the top of the mountain used to be now sat a huge horseshoe-shaped crater, and high mounds of landslide debris lay around the base of the volcano. Since Loo-Wit erupted in this unexpected fashion, volcano doctors (officially called volcanologists) have discovered that there are such craters and landslide mounds associated with many other composite volcanoes around the world—Mt. Shasta, in California, and White Island, in New Zealand, happen to be two of them.

The other devil of teapot volcanoes is mud. Take water from melting snow or ice or heavy rains, add ash squared, and you get mud to the fifth power. Great landslides, called mudflows, can race down the mountain during eruptions, or hundreds of years after a volcano's last eruption.

Mudflows from Mount St. Helens choked the Toutle River and went all the way to the Columbia. In A.D. 79, mudflows off of Mt. Vesuvius buried the town of Herculaneum, and in 1985, mudflows from Nevado del Ruiz in Columbia killed 23,000 people. Geologists have also identified no less than 76 mudflows from Mt. Rainier over the past 10,000 years. Some of these traveled more than 70 miles (113 km), all the way to Puget Sound near Seattle.

The end stage of violent eruptions from teapot volcanoes is the formation of sticky lava domes, which grow in the crater blasted out during a volent eruption. The lava dome growing in the horseshoe-shaped crater at Mount St. Helens is a typical example.

Bathtubs

"Nothing, not even the wind that blows, is so unstable as the crust of this earth," wrote Charles Darwin, which brings us to bathtub volcanoes, those Atlases of volcanic eruptions, those crust movers and hole makers. Bathtub volcanoes are the stickiest, nastiest, gassiest of the lot. Their eruptions are earth shaking and cataclysmic, often exceeding the power of many nuclear bombs. They are the bringers of famine, tidal waves, fire, pestilence, plague, and pretty awesome sunsets—death with a sickle and a rainbow.

These crust levelers erupt so much lava so fast and so violently that they create a great underground cavern, an empty space at the top of the magma chamber where the blown-out lava used to reside. Blowing out so much material so fast leaves the overlying rocks with no support. With nothing below them but empty space the rocks collapse into the cavern, forming a giant hole that geologists call a caldera. These volcanoes are thus called "big-hole-in-the-ground volcanoes," or on a simpler note, caldera volcanoes.

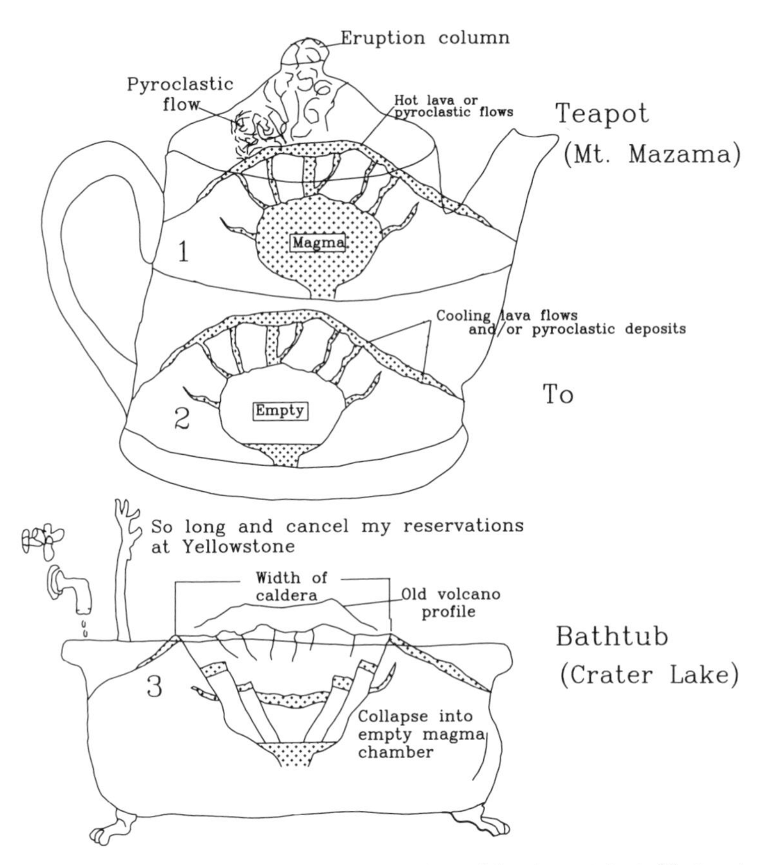

A teapot volcano changing into a bathtub volcano. So much lava is erupted out of the teapot over such a short time (1) that the top part of the magma chamber becomes vacant (2). With nothing below for support, the top of the teapot collapses into the empty part of the magma chamber, and a great hole, called a caldera, is formed (3).

Calderas come in two sizes—big and bigger. Big calderas, the blue whales of volcanic holes, have diameters measured in single miles; bigger calderas, the Seismosaurs of volcanic holes, have diameters measured in tens of miles. Crater Lake (6 miles wide, or

10 km) would be a blue whale, whereas Yellowstone (28 × 46 miles [45 × 74 km]) and California's Long Valley (25 × 12 miles [40 × 19 km]) would be Seismosaurs.

Most of the volcanic gas and lava bits blasted to kingdom come return to earth as either fallout ash or glowing avalanches (pyroclastic flows). However, enough fine material will remain up in the stratosphere to make pretty sunsets and to wreck havoc on outdoor picnics and winter heating bills for years to come.

The fallout ash can be tens of feet thick and can be delivered by the wind hundreds of miles from the eruption. The eruption of Yellowstone 600,000 years ago spread a layer of ash as thick as a Big Mac as far as Texas and Kansas. The eruption of Tambora on the island of Sumatra in 1815 left ash deposits over 20 inches thick (51 cm) up to 30 miles away (48 km) and blew pieces of pumice 6 inches wide (15 cm) over this same distance. Krakatoa blew ash 60 miles (97 km) into the air and left 1 foot of ash more than 70 miles (113 km) away.

Then there are the glowing avalanches, the sports cars of caldera eruptions. These can travel more than 100 miles away (161 km) from the collapsing volcano at speeds in excess of 150 miles per hour (242 kmph). At Crater Lake they went over 30 miles (48 km), carrying pieces of pumice up to 6 feet wide (1.8 m). The Long Valley caldera in California erupted some 700,000 years ago, sending the Bishop Tuff more than 30 miles (48 km) away from home. At Lake Taupo in New Zealand, pyroclastic flows, from a great eruption 20,000 years ago, traveled more than 65 miles (105 km) and managed to climb 1,000-foot (305-m) mountains to get there.

What about the great hole, the abyss itself? Like a landfill, it slowly fills up. It can contain some of its own eruptive debris (pyroclastic flows over 2,500 feet thick [762 m] have been found inside some calderas), or it can be filled by later volcanic deposits or by material falling or washing in from the crater rim. Most commonly, however, it is filled by rainwater to form a very pretty lake, as in the case of Crater Lake in Oregon.

Bathtubs are the ultimate in volcanic hazards, and they come well disguised; there are those that hide their true identity (the

Clark Kents) and those that get you when you least expect it (the Benedict Arnolds). Mild-mannered Clark Kent. No one knew what strength and power hid beneath his gray suit. The same is true for this type of caldera. There is no great mountain or volcanic hill present to tell of past eruptions or warn of future disasters. The volcanologist George Walker has called these calderas "negative volcanoes"—ones that have no topographic expression. The hot, sticky lava simply climbs aboard an earth elevator and heads for the surface. As it rises upward it pushes at the overlying crust until the crust bends and cracks, creating fissures that become highways to the sky for the gas-rich lava, the lava comes charging up the break like a space shuttle headed into orbit.

The first explorers to arrive at Yellowstone and see the hot springs, mudpools, and geysers thought they had stumbled through the gates of hell directly into Dante's *Inferno*. The devil himself walked the land 600,000 years ago. There was no tall mountain, no hill, the ground simply shuddered and cracked and all hell broke loose. Forests were incinerated and buried, ash fell to thicknesses of several feet over a distance of 50 miles (81 km), and glowing avalanches raced tens of miles to bury the landscape below hundreds of feet of hot, broken lava. The resulting hole measured 28 × 45 miles (45 × 74 km) and was well over 3,000 feet deep (914 m). Geologists say Yellowstone erupts this way every 600,000 to 800,000 years. It's now been 600,000 years since the last eruption. Are you planning a trip soon?

If so, as you travel across Montana and Wyoming enjoying the scenic views, imagine what such a volcanic eruption might be like if it occurred today, an eruption on a scale not known in recorded history.

It might start with a swarm of violent earthquakes breaking the ground open like a watermelon dropped from a 10-story building. A colossal explosion follows, and Old Faithful awakens to find out it never has to be on time again. Old Faithful, along with Yellowstone Lake and everything in between, is blown 30 miles (48 km) into the sky. Obsidian Cliff, Mammoth Hot Spring, Inspiration Point, and Tower Falls vanish as though they never existed, and windows as far away as Cheyenne, Salt Lake City, and Missoula are shattered like fine crystal broken by the high note of an opera singer.

Out of the open cracks for 800 square miles (2,072 km^2), hot, gassy lava explodes. The pulverized magma rockets skyward to form mushroom-shaped clouds, finger-like pillars, and dancing curtains that reach right to the stratosphere. Within minutes a bright, sunny morning has been turned into the darkness of a cave, a darkness that will remain for day after long, suffering day.

The explosions come one after another, a continuous barrage that makes a Rolling Stones' concert sound like wind chimes in a hurricane. With the explosions comes rain. A steady downpour, not of water, but of volcanic ash, sand, and gravel falling from the clouds and pillars and curtains, inch after foot after yard, to blanket and bury large parts of Montana, Wyoming, Idaho, Utah, Colorado, South Dakota, and the southern parts of Saskatchewan, Alberta, and British Columbia.

Here and there the blackness is broken by ragged bolts of green and red lightning and by dozens and dozens of giant, blood-red fireballs streaking cometlike through the sky as they plunge to the ground to become glowing avalanches; pyroclastic flows that incinerate forests and ranch lands, consume lakes and rivers, and destroy cities and towns. Such a disaster would be unlike any ever experienced in the United States.

So, when you think of Yellowstone, one of our great national parks, with its hot springs and geysers, streams and mountains, and bison, elk, grizzly bears, and wolves, also think of it as part of the corkscrew of time, a place with a violent geologic past and a certain, explosive future—a present-day time bomb ticking away in the Rockies.

There is another time bomb located some 550 miles (885 km) southwest of Yellowstone. Standing on the ski slopes at popular Mammoth, California, you would hardly think the peaceful valley you gaze into is slowly ticking away. Long Valley, a 19-mile-long (31 km) time bomb last erupted 700,000 years ago, sending pyroclastic flows more than 30 miles (48 km) to the southwest and northeast. The volume of material blown out was great enough to cause the ground to collapse 1.2 miles (2 km). Over the past 20 years the floor of Long Valley, including popular Mammoth Lakes, has risen 9 inches (23 cm), which, in geologic terms, makes it an elevator on its way to a penthouse party. This worrisome fact, which strongly indicates hot magma is rising toward the surface, coupled with the occurrence of numerous earthquakes in the area, led the United States Geological Survey to declare the area a potential volcanic hazard. Much like Captain Hook, I can plainly hear the tick, tick, ticking; how about you?

Other Clark Kent volcanoes include Taupo in New Zealand, which erupted 20,000 years ago, and Toba in Sumatra.

The Toba eruption, some 75,000 years ago, may have been one of the greatest volcanic explosions on earth. The resultant collapse formed a fishing hole some 30 × 60 miles (48 × 97 km), and the blown-out lava bits buried a 10,000-square mile (25,900 km^2) area in ash and pyroclastic flow material to a depth of 1,000 feet (300 m)!

Benedict Arnold volcanoes are composite volcanoes through and through until, when your back's turned, when you least expect it, they betray you to planet Earth. There are many Benedict Arnold volcanoes, but two come immediately to mind—Mt. Mazama and Krakatoa.

Snow-capped and majestic, standing 13,000 feet (3,962 m) above sea level, Mt. Mazama (now called Crater Lake) was a top-of-the-line Cascade strato-volcano. Some 7,000 years ago it erupted. No one knows why. It was just one of those lazy, crazy summer evenings. Restless and hot, the earth took a turn for the worse; "The ground started to shake. The rumbling noises grew louder and peals of thunder echoed over the mountain, interspersed with sharp detonations like the discharge of heavy guns."[4] The destruction had begun.

Powerful explosions sent ash over an area of 444,000 square miles (1,137,000 km^2), from British Columbia to Alberta and Montana. Up to 70 miles (113 km) away, over 6 inches (15 cm) of ash fell, and around the volcano it was more than 20 feet deep (6 m). Glowing avalanches raced over 37 miles (60 km) away, carrying pieces of pumice more than 6 feet wide (1.8 m). Ice and snow melted, causing great mudslides. Darkness was everywhere.

When the air cleared there was no mountain. Instead, a boiling, seething ragged crater some 6 miles wide (10 km) and 4,000 feet deep (1,219 m) sat in its place—surely this was the throne of Satan. Then the rains came; steam rolled in long columns to the sky, and the crater filled slowly to form the deepest freshwater lake in North America—Crater Lake.

Then there was the shot heard round the world—not the one at Concord, Massachusetts, but at Krakatoa, west of Java. A hill of humble beginnings, Krakatoa was now a peak only 1,500 feet (458 m) above the hungry sea. Krakatoa had been idle for more than 200 years, a curiosity on the sailing charts of steamships, until 1883.

Imagine being a cabin boy aboard a coal ship in the Sunda Straits on August 26, 1883:

> There was an explosion sounding like the earth had split wide open. First I thought it was our boilers but looking toward the island of Krakatoa I saw a huge, black mass rolling into the sky. It began to advance towards our ship, lightning shot out all sides and the rolling thunder was like a great war ship bearing down on us, guns blazing. We were certain we were doomed. Then the showers started; not rain to wash the dark cloud away but sand and coarse gravel. It fell like hail and we had to hide where we could. Every few minutes the captain would shout and we would run out to sweep the heavy sand from our decks.
>
> Then came more explosions, louder than the first. So loud they shattered the ear drums of half our crew. All was blackness. The wind picked up as had the sea. We rolled from side to side, the terrible rain falling and blackness all around-surely the day of judgment was upon us.

The eruption lasted two days. The loudest explosion was heard more than 3,000 miles (4,800 km) away. Ash was blown 50 miles (81 km) into the air, and windows were shattered more than 93 miles (150 km) away. For hunderds of miles around day became night, and on the island of Sumatra the night lasted for 2½ days. It is estimated that 7 square miles (18 km^2) of material was erupted at a force equal to a few thousand megatons of atomic bombs. Pumice rained out of the air over hundreds of miles and glowing avalanches inundated the Sunda Straits. Krakatoa collapsed, leaving a hole 3 × 6 miles (5 × 10 km) with a depth of 3,000 feet (914 m). The collapse caused tidal waves that swept across the sea and over the islands of Sumatra and Java. These were over 100 feet high (305 m) and went more than 10 miles (16 km) inland, killing more than 36,000 people.

There was so much pumice blown out that it filled the Sunda Straits up to 10 feet deep (3 m). Some pieces were big enough to carry human bodies, which one did for 3,825 miles (6,158 km) across the Indian Ocean to Africa. Pumice littered the sea 1,250 miles (2,012 km) away from Krakatoa and, at times, was so thick that sailors could get off their ships and walk around on it. Most pieces were brick size and provided great homes for crabs and other small sea creatures.

Then there were those Krakatoa sunsets.

> But who shall paint the glory of the heavens when flecked with clouds?—burnished gold, copper, brass, silver such as Turner in his wildest dreams never saw, and of such fantastic forms. (E. L. Layard of New Caledonia, some 4,203 miles (6,767 km) east of Krakatoa)[5]

> Last Sunday, about 5 o'clock the sun looked like a blue globe . . . and after dark we thought there was a fire in the town from the brightness of the heavens. (News report from Trinidad, September 2, 1883)[6]

People thinking their town was on fire was quite common. In Poughkeepsie, New York, the fire department was called out and the fire bells were rung because of the extreme red illumination in the southwestern part of the sky. This was more than three months after the eruption.

It is estimated that Krakatoa's ash reduced solar radiation by 10% for a period of 3 years. Needless to say the summers were cool, and the winters downright frigid.

Eruptions like Krakatoa occur about once or twice a century; they are small compared to a Yellowstone or a Long Valley, and so the earth turns and humans continue to build around places meant for dragons, wizards, and space suit–clad scientists called volcanologists, who are up on the mountain trying to practice preventive medicine. These scientists monitor volcanoes with a variety of instruments from lasers and heat probes to tiltmeters, seismographs, and gas analysis. They hope to use the data from such studies to forecast when and how a volcano can or will erupt.

The need to know is great in terms of life and property. But we learn only by observing and waiting, testing and waiting, watching and waiting. "Nature is often hidden, sometimes overcome, seldom extinguished" (Francis Bacon). Seems like each and every darn volcano has a mind of its own, its own habits and lifestyle. If that is indeed true, then we have to monitor all volcanoes that people live around. Dormant today or for a million yesterdays doesn't mean dormant tomorrow. The life span of a volcano is much longer than collective human memory.

So volcanologists are up on the angry mountain, risking and sometimes losing their lives, as they have at places called Unzen and St. Helens. The volcano doctors have come a long way in forecasting eruptive behavior, but they still have miles to go before they nap, let alone sleep.

ASHES TO ASHES, DUST TO US

Ashes to ashes, dust to us,
Volcanoes cause civilizations to rust,
Just like a car
They come and go,
Leaving only the volcano to know.

Earthquakes and volcanic eruptions have had a devastating effect upon people, civilizations, and possibly the march of history.

For instance, in the third century A.D., the Mayas of Central America had developed quite a sophisticated society. They were interested in science and medicine, they had great trade routes, and they had built large cities; their civilization was clearly flourishing.

Then disaster struck. The volcano Ilopango erupted, leaving an area up to 60 miles (97 km) around it completely uninhabitable. At a village called Chalchuapa, some 50 miles (80 km) from the volcano, ash over 2 feet deep buried the houses, fields, and crops. The ash layer became thicker the closer one got to the volcano, until it was 160 feet thick (49 m) at the monster's base. Entire forests were uprooted and incinerated and thousands died instantly. All crops and farmland were destroyed or under many feet of volcanic ash, and famine ruled the land. The survivors fled, and so ended Mayan civilization in what is now El Salvador. The survivors fled to Guatemala and what had been small, insignificant settlements there turned into centers of Mayan civilization. One of these was the city of Tikal, which, at one time, had more than 50,000 inhabitants—all thanks to a volcano in El Salvador.

In A.D. 1100, the great Olmec-Mayan city of Teotihuacán was larger than the walled city of Imperial Rome. With a population of more than 200,000, it was a great center of art, commerce, science, and religion. It had a vast, wondrous marketplace where one could buy or barter for pumpkins, corn, chocolate, squash, beans, peppers, potatoes, pigs, turkeys, avocados, coca, beer, and who knows what else.

Overnight this great city was virtually destroyed. Archeological excavations uncovered evidence of a great fire, collapsed buildings, and skeletons in sitting and reclined positions when they were struck. Rising up behind the great city, Cerro Gordo ("fat mountain") erupted violently and sent hot, fast-moving pyroclastic flows down its flank, and suddenly the city was no more. Aztec stories tell it this way: "Time ended when the volcanic eruptions occurred. Fire and cinder rained down out of the sky and consumed the earth."

In 1815, the greatest eruption in modern history took place at Tambora on the island of Sumbawa. Tambora vanished, as did 6,864 feet (2,092 m) of volcano; 12,000 were immediately killed, and another 85,000 died from tidal waves and the resulting famine. Zillions of bits of lava were hurled into the stratosphere to encircle the globe. The summer of 1816 was a disaster in North America and

Europe; cold and rain and crops dying or dead. Temperatures hovered around 50° F (10° C) for daily highs in the eastern United States and Canada, and they dipped below freezing at night. A blizzard on June 6 dumped 6 inches (15 cm) of snow across New England, and on July 4 the temperature in Georgia reached a high of only 48° F (8.9° C). In Europe crops failed and the resulting famine was so severe that people were forced to eat moss, and there were food riots in France. This was also the year that started the exodus of New England farmers west to the Ohio country and Illinois and Indiana. What if the volcano had not erupted—how then would the West have been won?

And we have all heard about the island of Thera, the eruption of Santorini, the lost city of Atlantis, and the Minoan civilization that ruled the Aegean Sea, which the eruption of Santorini in 1500 B.C. put an end to. Santorini was one of those Benedict Arnold volcanoes, and when it decided to erupt it went the whole nine yards. Santorini, like Krakatoa, collapsed into the sea, sending great tidal waves across the Aegean. These walls of water were more than 200 feet high (61 m). The Minoan cultural and economic center on Crete was ruined, and other Minoan towns along the coast were shattered and destroyed; thus a great civilization was so terribly weakened that it became easy prey for others. However, the eruption that destroyed the Minoans may also have saved the children of Israel. The eruption occurred about the time Moses and his people were fleeing the Egyptian soldiers. The Red Sea parted for Moses but swept the Egyptians away. It is not uncommon for water levels to drop tremendously just prior to the coming of a great tidal wave. A civilization destroyed, the legend of Atlantis alive, and the children of Israel safe and sound—possibly all because a volcano went kaboom—and the earth plays on.

Prelude to Chapter 3

A grinding crack and a fault fails. The earth rumbles and shakes. Dishes, pictures, and lights crash to the ground; bridges, freeways, and buildings break, crack, and topple. And the music? The low, steady hum of the double basses as the earthquake rumbles across the land. Occasionally, heard above the basses, are sharp, slashing chords from the violins as things crash, tumble, and break. This music continues, on and on, until it comes to an abrupt end.

CHAPTER 3

Living on an Eggshell

'Tis Earthquake: brother to the volcano and the sister of death.

I knew the frantic EARTHQUAKE in his car
had rattled by, and laughed; and visions swift
Trooped o'er my brain, of horrors manifold
That have befallen when this mighty orb
Cracked like a globe of glass, alarming nations
With the wild thunder; whose deep rung vibrations
Ran jarring from the tropic to the pole:
When cities shook, unseated; and loose walls
And staggering towers across the peopled streets,
Nodded and knocked their heads in ponderous ruin
Deep-burying all below; wildest convulsions
of all that agitate the frame of nature!

—FLACCUS

For no apparent reason the dogs started to bark and whine; the smallest one trembled. The cat began yowling and soon after started to run in wild circles.

Then came a deep rumbling, like summer thunder, and a great whooshing noise, as if thousands of locusts had invaded the trees. Out of nowhere the ground rose up and knocked me flat. It rippled and swayed; trees rose and fell like logs on a windy lake. Parked cars seemed to be going two ways at once, rocking violently from side to side while trying to move backward and forward. Telephone poles leaned left then right, some so far they almost touched the ground; others spun like dizzy tops.

Helpless, I hugged the ground, that rock-solid earth that now rippled and rolled to the music of a rumbling earthquake. Then, as quickly as it came, it ended. This was followed by a long moment of quiet and calm before the sirens and screams.

In Indonesian mythology the great snake Hontobago had wiggled and coiled its great body; in Japanese mythology the giant catfish, on whose back the world rides, had wiggled its fins. Or, possibly, it was the work of Loki the trickster of Norse legend. Long ago Loki, out of pure spite, arranged the death of Balder, Odin's son. For punishment Odin had Loki tied to three giant stones in an underground cave. A great snake was hung above Loki's face, its venom dripping straight into his eyes. Loki's wife stood beside him holding out a goblet to catch the poison, drop by drop. When she turned to empty the goblet, a drop would strike Loki's eyes and he would twist and thrash and the earth would tremble and shake.

Loki's screams were heard the day Joshua stood pondering how to overcome the thick, tall walls of the city of Jericho. Suddenly, from nowhere, it came "to pass, when people heard the sound of the trumpet, and the people shouted with a great shout, that the walls fell down flat, so that the people went up into the city . . . and Joshua won the battle of Jericho."

Archaeologists, excavating part of the old city, have found walls 13 feet (4 m) thick and 30 feet high (9.1 m)—way too massive for a trumpet (even Louis Armstrong's) and a yell (even Geronimo's) to have toppled them. Other evidence from the excavation shows there was a great fire, and the way the walls collapsed (like maids, all in a row) strongly suggests an earthquake.

SHAKE, RATTLE, AND ROLL

Earthquakes cause the ground to shake, rattle, and roll. But where do earthquakes reside, and how do they reach out to squeeze the land? Earthquakes usually hang out downtown right where all the action is. Down along the edges of jigsaw pieces, where plate scrapes and grinds against plate to let loose the deadly forces that shake the ground and destroy cities.

Crack a walnut. Sometimes, with almost no effort, it falls apart. Other times you can squeeze and squeeze the nutcracker and nothing happens until, with a sudden crack, the walnut explodes in tiny pieces. Imagine being 40 miles (64 km) below the surface of the earth, right at the edge of a jigsaw piece (of course you are hair thin because the great pressure down there has squeezed you as flat as a stick of gum). You have a front-row seat to watch as plate pushes and shoves against plate, brittle rock squeezing brittle rock like a nutcracker squeezing a walnut. The pressure on the rocks builds and builds until one rock cracks and slips against the other. The rock has broken and a geologic fault has formed. A fault is a crack along which there has been displacement of the sides of the crack relative to one another. Not my fault, not your fault, but the earth's fault.

This cracking, or faulting, releases stored-up energy, causing the ground to quiver like a tuning fork and to vibrate like the air around a ringing bell. These vibrations race away from the cracking point (called an earthquake focus), like waves on a windy lake. Rolling through the earth these seismic waves cause the rocks and soil to ripple and undulate, creating the fearful "quaking of the earth."

After this initial snap and the resulting earthquake it takes the broken rocks some time to adjust to not being pushed around, during which time numerous small earthquakes called aftershocks occur. Though not very powerful, they can still do significant damage to buildings already weakened by a major quake. During the 4-month period after the 1964 Alaskan earthquake, there were more than 1,200 aftershocks. Most earthquakes happen within 60

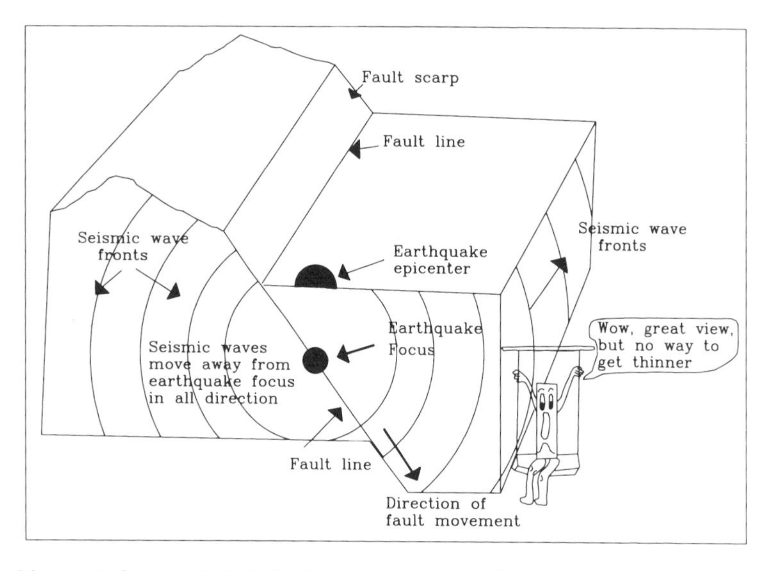

Movement along a geologic fault releases energy that travels away from the cracking point (earthquake focus) like waves on a lake. Moving through the earth's crust these seismic waves cause rocks and soil (and anything built into or upon these) to vibrate and jiggle. (Adapted from C. Montgomery, and D. Dathe, Earth Then and Now, *1991)*

miles (97 km) of the earth's surface—right in the middle of our piecrust (oceanic crust) or whipped cream (continental crust). They take place here because this is the most rigid part of the earth, the part capable of being cracked like an eggshell. Relatively few earthquakes take place within the pie filling (asthenosphere) because it is dominantly composed of hot, soft rocks. This material can be bent and twisted like a cruller, but still it won't break. So the only places that earthquakes occur in this soft, runny stuff is where a piece of the rigid crust is pushed down into it.

Earthquakes hibernate, waiting for wake-up calls, where continent meets continent—India pushing into Asia, Africa shouldering into Europe. The Chinese have had so many devastating earthquakes that 2,000 years ago they started the science of seismology (study of earthquake waves) to try to determine the direction to the source of an earthquake. The Chinese invented the first earthquake sensor in the second century A.D. This was in the form of a large, ceramic urn with eight dragon heads around the outside. During an earthquake, a swinging weight inside the urn would cause one of the eight dragons to open its mouth. With a loud clang a small brass ball would drop from the dragon's mouth into a metal frog attached to the urn directly below the dragon. The dragon with the empty mouth indicated the direction to the source of the earthquake.

Earthquakes also sleep lightly where plate bends below plate, or where one plate slides against another plate. Such a place is a wiggly line some 600 miles long (950 km) called the San Andreas fault. This is where the North American plate slips by the Pacific plate, carrying a piece of the West Coast steadily northward. San Diego will pass Oakland about 15 million years from now, and in the last 1 million years, Los Angeles has moved 31 miles (50 km) closer to San Francisco. Most of the movement along the San Andreas fault (and most other faults) occurs in fits and starts, not in constant sliding. The fits and starts are the rocks snapping under decades of pressure to let loose earthquakes and, in one shot, move Los Angeles 10 feet (3 m) closer to San Francisco, to the chagrin of both those in the Bay Area and Los Angles.

Finally, earthquakes can occur out in the middle of nowhere, geologically speaking of course, in places far from the action, such as in the heartland of America, in the prairies of Canada, or in the middle of Australia.

Or in places like New Madrid in the winter of 1811–1812. This was Missouri Territory, or—as Davy Crockett called it—shakes country. Three big earthquakes struck close to the place where Kentucky, Tennessee, and Missouri meet. The ground shook from New Orleans to southern Canada, and from New England to Georgia; in fact, the quakes rang the church bells in Boston, some 1,000 miles (1,600 km) away. The Mississippi River changed its course, forests were flattened, and towns were filled by flooding rivers. Lakes were lifted "up into the sky" and emptied, while in other places the ground sank enough to create new lakes (such as

St. Francis and Reelfoot in northwestern Tennessee). An estimated 150,000 acres of timberland were destroyed, and aftershocks occurred for 10 years. Imagine such an earthquake in this area today, an area occupied by more than 13 million unprepared people.

The New Madrid earthquakes occurred along old cracks, cracks that may have formed when the North American continent tried to split apart some 1.1 billion years ago. No one really knows for certain the reason why Missouri Territory got the shakes. Possibly it was the North American plate relieving some of the stress its under along old structures thousands of miles from where the action is. Or possibly the Midcontinent Rift isn't dead, only slumbering and tossing and turning in its sleep as it waits for the wake-up call. There have been numerous tremors in this same area between 1974 and 1989 (most less than 3.0 on the Richter scale).

Of more importance is the recurrence interval for large earthquakes (greater than 6.0) in this same area. Based on the rate at which recent, mild earthquakes occur, and the age dating of faults produced by earthquake events in the area, it is generally accepted that a large earthquake will occur every 100 years. The last earthquake of magnitude 6.0 or more occurred in 1895 near Charleston, Missouri. So, I'm not trying to "shake" any of you 13 million up, but I never realized Missouri, Kentucky, and Tennessee had so much in common with California.

This earthquake recurrence interval of 100 years led a self-professed climatologist by the name of Iben Browning to predict (with a 50% probability) that a large earthquake would occur in the New Madrid area on December 2 or 3, 1990. This prediction became credible, as well as a good story, to many newspaper and television reporters. As a result, the Browning prediction caused much public concern. Needless to say, nothing happened but the hue and cry, which would have probably registered a magnitude 7.0 or more, led the United States Geological Survey to publish a circular entitled "Responses to Iben Browning's Prediction of a 1990 New Madrid, Missouri, Earthquake."

Earthquakes are the brothers to volcanoes. There are several small earthquakes per day recorded by seismographs around historically active volcanoes when they are supposedly dormant; during eruptive phases the number can climb to hundreds per day. Most volcanic earthquakes are caused by the movement of lava or volcanic gas below or within the volcano. As this material moves, the surrounding rocks are broken and cracked open by the drastic changes in temperature and pressure. Most of these earthquakes are relatively mild, but they can lead to disastrous results. It was an earthquake on the morning of May 18, 1980, that started the great landslide that led to the resulting explosive eruption of Mount St. Helens.

HOW BIG IS BIG?

Everyone who lives in earthquake country talks about the coming of the really big one, the granddaddy of all quakes. But like fishermen and the size of their fish, how big is really big? How do we measure the size of one earthquake against that of the next? Geologists have two different scales that can be used to determine which earthquakes are related to Godzilla and which to puppy dogs.

The first is based on what is called earthquake intensity—the measure of the effects of an earthquake on you, me, and our buildings. In 1902 an Italian physicist by the name of Giuseppe Mercalli came up with a scale based on the amount of damage an earthquake caused to various types of structures. Mercalli varied his scale depending on the distance from what is called the earthquake's epicenter—the point on the surface of the earth directly above the source of the earthquake—and the strength of the earthquake. A 1 or 2 on Mercalli's scale means an earthquake is barely perceptible to people; a 5 or 6 means windows and dishes are broken, furniture moves, bells ring, and trees shake. A whopping 10 would leave buildings flat and cause major landslides, and a 12—you and everything around you are history.

The other way scientists like to rank earthquakes is by what is called magnitude—the total amount of energy released or the strength of the quake at its source. This is just another way of figuring out how much the ground shakes.

In 1935 Charles Richter, a geophysicist at the University of California at Berkeley, devised a way to rank earthquakes based on the amount of ground motion as registered by a seismograph, an instrument that uses a pendulum (like the one Poe used in the pit) to record the vibrations of the ground on a moving strip of paper. The greater the motion of the pendulum on the paper, the bigger the earthquake and the greater the magnitude.

Today the Richter scale is used worldwide to describe the bigness of an earthquake. The magnitude is determined by measuring the maximum amount of displacement of the pendulum relative to the paper during an earthquake. Because seismic waves get weaker the farther they travel from their source, adjustments must be made for the distance the seismograph is from the earthquake focus.

The largest earthquake ever recorded on the Richter scale (which has no upper limit) was 8.7—released energy equal to the detonation of 1 billion tons of TNT. An earthquake less than 2.5 on the Richter scale is not felt by people. The Richter scale is logarithmic, since the ground-shaking ability of earthquakes varies by orders of magnitude. Thus, an earthquake of magnitude 5 on the Richter scale is 10 times greater than an earthquake of magnitude 4, and 100 times that of a 3.

The real difference between a large and small earthquake, Godzilla or a puppy dog, is related to the total amount of rock that is under stress at any given time. Five miles of breaking, slipping rock equals less than a magnitude 2.5 and is a tail wagger; 300 miles of rock crushing rock and we have a magnitude 7 or more, and a new monster is born.

There are over 1 million earthquakes every year recorded by seismographs around the world. The vast majority of these are less than 2.5 on the Richter scale (softer than a dachshund running across a carpeted floor). There are about 800 damaging earthquakes a year (5.0–5.9, about equivalent to a Saint Bernard pounding across a wooden floor), 120 destructive quakes (6.0–6.9, or a bull in a china shop), 18 major quakes (7.0–7.9, or King Kong chasing the bull in a china shop) and 1 or 2 great earthquakes (>8.0, or Godzilla after King Kong, who's still chasing that bull). On average more than 10,000 people die each year from earthquakes.

In the past few years earthquake scientists have started to abandon the Richter scale and use instead what is called the "moment of magnitude" to measure earthquake intensity. The reason for this is that the Richter scale is apparently not accurate when it comes to measuring the big ones (>8) or when the epicenter is more than 100 miles from a seismograph. This is not a problem in California, where there is one seismograph for every limousine, but in places like China and India it most definitely is.

Moment of magnitude is based on the size of the fault on which the earthquake occurs, and the amount of movement or jump that takes place along the fault. The larger the fault and the greater the movement, the bigger the moment of magnitude.

Moment of magnitude numbers are routinely translated into Richter numbers.

THE MOVERS AND SHAKERS

Earthquakes are movers and shakers and breakers; they are the CEOs of the planet. Seismic waves cause the ground to vibrate up and down and to ripple from side to side. This is not good if you are an object that sits on, or is built into, the ground. Buildings, roads, bridges, and power and telephone poles collapse, crack, break, splinter, or explode. Wooden structures and new types of concrete, which are reinforced with steel, have the ability to flex and give, and thus possibly survive. In fact, there are even "shock absorbers" that can be built into the foundations of new buildings. Theoretically, during an earthquake, the buildings will ride up and down with all the smoothness of a Cadillac on a bumpy, dirt road. Brick, mortar, and old concrete, on the other hand, cracks, breaks, and collapses.

Landfill, flood plains, wetlands, and sand and silt can liquefy when vibrated. The soil or sand or silt particles are shaken violently apart, allowing water to seep between them to make quicksand. Because of this they end up shaking like Santa Claus' belly.

In 1556, in the Shensi region of China, an estimated 830,000 people died when an earthquake struck on January 24 at 5 a.m. Many of the victims lived in homes carved out of glacial clay. When the ground shook and rippled the walls of the houses liquefied and collapsed. The falling roofs buried the sleeping people. Many of those who did survive the quake later died from the resulting famine.

The earthquake that struck Mexico City in 1985 sent ripples and quivers through the soft landfill material upon which part of the city is built. This mushy stuff amplified the earthquake vibrations 5 times more than would have good old solid rock. Houses built on the landfill ended up looking "like a giant foot had stepped on them."

Then there was Umarga, India, in the fall of 1993. It was early morning and all the world was asleep when a 6.5 magnitude earthquake came to visit southwest India. The mud-and-mortar homes rippled, liquefied, and collapsed on top of the sleeping families. Some 30,000 people died, and entire villages ended up looking like Godzilla had been visited on them.

Loose upon the land, Godzilla paid a visit in June of 1995 to the town of Neftegorsk, located on Sakhalin Island, Russia. A 7.5 magnitude earthquake liquefied the sandy island, and the apartment buildings in the town "collapsed like a house of cards," killing over 2,000 people. Neftegorsk was an oil-drilling town built during one of the Soviet Union's grand five-year plans. Between 1967 and 1972, 17 five-story apartment buildings were thrown up on the sandy soil. The apartments were boxes constructed from 2-ton panels of prefabricated, unreinforced concrete. All 17 of the apartments collapsed like accordions. "When the sun sets and the rescue equipment stops, the town starts moaning. These are the moans of those buried under the rubble" (*U. S. News and World Report*, June 12, 1995, p. 14). The only things left standing were single-story wooden homes and a 22-foot (6.7 m) statue of Vladimir Lenin. The statue was the tallest structure in town to remain upright; obviously it was the only thing the old Soviet Union built that was meant to last.

Stopping the World Series, the 1989 Loma Prieta quake registered 6.9 on the Richter scale, and thus became the strongest quake along the San Andreas fault since the 1906 San Francisco quake. Though deaths were few, property damage was approximately 6.5 billion dollars. Much of this was due to the collapse and breaking of old concrete buildings and bridges, some of which, like the Nimitz freeway in Oakland, had been reinforced. Moreover, houses built on the jelly of the marina area of San Francisco completely collapsed; again the earth had issued a warning about the dangers of landfill and mud and mortar in quake country. Think anyone was listening?

Five years later, at 4:30 on the morning of January 17, a 6.6 earthquake rumbled across the San Fernando Valley of California. When it was finished, the so-called Northridge quake had killed 50 people, injured 6,000, and left over 20,000 homeless. Six major freeways, all made out of concrete, broke, buckled, and snapped; hundreds of houses collapsed or were badly sagging, and many businesses were simply out of business. The quake caused 30 billion dollars worth of damage, and the aftershocks were frightening. The earthquake may also have started a plague. Apparently, the earth shook so hard that bacterial spores in the soil were released into the air, leading to an outbreak of what is called "valley disease" and several deaths. Two major earthquakes in 5 years, neither even close to the Big One, but possibly they represent the opening credits in that great epic of how the west will be lost.

Earthquakes also give rise to great walls of rushing, roaring, killing water called seismic sea waves, or tsunamis. Tidal wave is the popular term for this phenomenon, but these waves have absolutely nothing to do with tides or tidal processes.

Tsunamis are caused by the shaking of the seafloor. This movement sends waves surging outward like ripples on a lake's surface caused by a thrown rock. Once formed these ripples can travel along at speeds of 300 to 600 miles per hour (483 to 966 kmph). Despite such speeds they can go virtually undetected in the open sea because here the waves are only 2 to 6 feet in height and have distances between wave crests of hundreds to thousands of feet. However, upon tearing into the shallow waters of coastal areas or harbors, the waves start to drag along the bottom and slow down. They begin to pile up like a carpet being pushed along the floor and into a wall. The waves pile up 60, 100, 140 feet high (18 to 42 m) to become great roaring walls of water. Usually, the first warning that something is rotten is the rapid withdrawal of water from shallow areas. Here you are with a great spot on the beach, right beside the water, and suddenly the sea is a mile away—all you see is wet sand. At that moment forget the cooler and start thinking like Carl Lewis, for you have between 5 and 60 minutes to get yourself higher than the incoming wave—higher than 130 feet (40 m) to be really safe. The power behind tidal waves can drive them 5–10 or more miles inland (8–16 km), leaving boats, houses, piers, and fish strewn about hill and dale.

Believe it or not, tsunamis formed by an earthquake on one side of the Pacific, can do damage on the other side of the ocean. A tsunami caused by a 1960 earthquake in Chile was still strong enough to cause 21-foot (6.4-m) breakers in Hawaii 15 hours later. In 1946 an earthquake under the Aleutian Islands sent seismic sea waves racing south across the Pacific Ocean. They came ashore at Hilo Bay in Hawaii, having traveled 2,000 miles (3,220 km) in 6 hours. Entering the bay the waves grew to a wall of water some 55 feet high (16.8 m). Sweeping inland they destroyed over 500 homes, killed 50 people, and caused 25 million dollars of property damage.

It was 1964 and Good Friday in Anchorage and Valdez, Alaska. The ground rolled and rippled for 5 to 7 minutes, destroying buildings, streets, and railroads. The magnitude of the earthquake was 8.5 to 8.7 on the Richter scale, one of the biggest quakes ever recorded. Big quakes give rise to tsunamis, and within minutes after the ground stopped shaking a wall of water over 80 feet high (29 m) pulverized Valdez, Chenega, Whittier, and Seward. This wave swept miles inland, carrying boats, houses, chickens, people, and death. This tsunami also caused 11 deaths and extensive property damage in Crescent City, California, some 1,550 miles (2,500 km) to the south.

Unfortunately, some of the things earthquakes knock over or break can blow up and/or start fires. On top of this earthquakes commonly crack water mains needed to fight the resulting fires.

The earthquake of 1923 centered near Tokyo and Yokohama killed 143,000 people. The earth shook and rumbled for 5 minutes and over half a million buildings collapsed. Then a giant wave rolled over the two cities, only to be followed by great fires that burned out of control for three days. These terrifying events gave truth to the Japanese proverb that lists the five main terrors of Japan as earthquakes, fires, thunder, floods, and fathers. It is beyond me how the Japanese could compare fathers to the other four—they must know something I've completely missed.

San Francisco was lively and free-wheeling in 1906 when the San Andreas fault decided to jog 18 feet (5.5 m). The ground responded with a good 40-second shaking, and people, animals, and buildings were knocked flat. In the harbor area, where sand had been used for landfill, the quake proceeded to shake everything to pieces. The collapsing buildings threw burning kerosene lamps to the ground and tipped hot wood stoves over—and the fires started. The ground shaking also broke water mains, leaving little water available to fight the fires. The fires, which lasted for three days and killed 700 people, burned 521 blocks of downtown San Francisco, including the biggest hotel in the world, the Palace.

Enrico Caruso, the greatest tenor in the world, was staying at the Palace and had just given a performance of the opera Carmen the night before (his voice was probably no match for that of the earthquake). After the earthquake he was out walking around the city and met the great American actor John Barrymore. Barrymore had been at a private party and was still dressed in his finest clothes. Caruso was amazed and amused at Barrymore's fancy dress; "Mr. Barrymore," he said, "you are the only man in the world who would dress for an earthquake."

In some Italian neighborhoods in the burning city, people fought the fire by pouring gallons of homemade wine onto their roofs to keep the shingles wet. It worked extremely well and the houses were saved. Of course, without wine things just weren't the same. One homeowner was heard to remark, "Should have saved my wine and let the damn house burn. The wine would have made my life calm and beautiful; instead I have this old shack which creaks and trembles filling my nights with terror and my days with hard work."

At noon on a warm June day in 1993 a 7.8 magnitude earthquake struck Okushiri, Japan. The largest earthquake to hit Japan in 15 years, it was powerful enough to move the entire island westward more than 6½ feet (2 m). Minutes after the ground settled, tidal waves over 18 feet high (5.5 m) crashed into the Japanese coast. Buildings, houses, and cars, some with people still in them, were swept into the sea. The waves rushed across the water to pound the Russian and North Korean coasts. Dozens of people were killed by the sea wave, but worse was to come. Since it was lunchtime many cooking stoves were burning. The earthquake tipped these over and they started fires that burned over 3,000 homes to the ground. "I never had much but what I had I lost it all"[1] was a common, but tragic statement of the coastal fishermen.

And among many people who live in Kobe, Japan. The earth "roared" for 20 seconds on January 17, 1995, and the "Great Southern Hyogo Earthquake," at a magnitude of 7.2, destroyed 56,000 buildings in the city and ruined freeways and railroads. It left 290,000 people homeless, killed more than 5,000, and injured 26,000. Repairing the material damage will take years and billions of dollars; repairing the psychological damage of a people who boasted of earthquake-proof buildings and freeways will take much longer. As one elderly survivor said, "It's really difficult for a person my age to think about what to do next. I think this is as much as we could endure."[2]

TELL IT TO THE MOUNTAIN

Everyone wants to know if scientists can predict when an earthquake will occur. Can they do it days or weeks or years ahead? Can they tell how big it is going to be, as in "run for your life" or "the chandelier might shake"? And can they predict where the earthquake will occur, as in town or county or fault system? Unfortunately, as of now, only the dark inside of the earth knows for sure.

But don't despair, there are geologists out there, working for you day and night. One of the newer approaches these scientists are trying is to study what has come to be called "precursor phenomenon." No, they are not trying to contact the spirit world to find out what the earth is up to. What these scientists are doing is studying the things that happen to the ground, to animals, and to rocks prior to earthquakes. These changes, or precursors, may occur months, weeks, days, or just hours before a major earthquake—which makes earthquake prediction about the same as walking on a razor's edge. Some of the possible precursors include:

1. Possible patterning or buildup of small earthquakes (foreshocks) before a major quake.
2. Swelling and/or small changes in elevation (ground swelling) prior to a major quake.
3. An increase in the concentration of certain gases (radon, chlorine, methane) in groundwater or well water. As the rocks push and grind, tiny cracks open in them, allowing water greater access to, and faster movement through, the rocks. This is much like building a four-lane highway through downtown when it was only two lanes before: increased water traffic can lead to increased concentrations of radon or other gases.
4. An increase in the occurrence of natural radio waves in the ground to see if there is any increase prior to a big earthquake. This was noticed just before the 1989 Loma Prieta earthquake but was missed in the Northridge quake.
5. Detection of any movement of rocks below the ground. A beam of laser light is sent from one side of a fault line to a reflector on the other side. By calculating the amount of time it takes the laser beam to bounce off of the reflector, scientists can determine very small ground movements along the fault.

6. Unusual behavior of animals (the four-legged kind) who may be able to sense a coming earthquake. Whether they detect small shivers of the ground, rising gas, or creaking rocks, no one knows. But animals have been known to show weird and strange behavior prior to an earthquake, including snakes coming out of hibernation to crawl out into the snow and freeze to death, frogs jumping out of lakes through the ice, dogs and cats acting crazily, earthworms coming up out of the soil like lemmings leaping into the sea, and pet birds battering themselves against the bars of their cages.

It was a cold and snowy February, 1975. There had been weeks of small earthquakes, increased amounts of radon gas found in the groundwater, and the animals were acting like Dr. Doolittle was coming to town. Some animals were even coming out of hibernation, just to freeze to death. Something was very much wrong.

Based on these precursors Chinese geologists made a bold forecast. They said there was a distinct possibility of a major earthquake near Haicheng in northeast China, and that it might be only days away. The government listened and evacuated several million people from their homes, moving them to open places. Nine and one half hours after the evacuation was completed, a great earthquake rumbled over the horizon shaking the ground like a million buffalo thundering across the open plains. Buildings collapsed and houses were flattened, but tens of thousands of lives were saved by the evacuation.

Luck? Good science? Tea leaves?

Over the next couple of years these same scientists, filled with the heady elixir of success, concluded that a big earthquake was due in T'ang Shan, a city located southwest of Bejing. The problem was that they didn't have a clue as to when. Half a loaf is better than none you say? Unfortunately, not to the 650,000 people who died when the earthquake came to town. A whopping 8.0 on the Richter scale and there were no precursor warnings—no small quakes, no radon gas, no ground swelling, and all the animals were sound asleep. Just what is a scientist to do?

Prelude to Chapter 4

It began with an ear-splitting crack, like a sharp rap on a side drum. The top part of the mountain visibly trembled, then slid away. The cracking sound was immediately followed by a deep roaring, loud and steady, but gathering volume, like tremolos in piano bass accompanied by trombone glissandi. Down the mountain the avalanche raced and, with a crash of cymbals, it buried the valley. Finally, on the last quivering notes of the trombone, dust filled the air and all was silent.

CHAPTER 4

And Away We Go

The hills are shadows, and they flow
From form to form, and nothing stands,
They melt like mist, the solid lands,
Like clouds they shape themselves and go.
—Alfred, Lord Tennyson

Hills on the go, continually eroded and moved down slope to their final resting place, down to the sea. This process can take longer than an English cricket match—an eternity to wash soil and rock to the ocean. Or it can occur faster than a tennis serve—a second for the mountain to slide and fall, to crash in a whirlwind of rock and dust onto an unsuspecting town.

Looking back I realize the animals were the first to know the mountain had decided to come to dinner. That very morning one of the ranchers told me his cows refused to graze on the grassy slopes below the peak. Later Julie, a local painter, came by to tell me of the fox and raccoon she'd seen from her studio. They came up her road, then turned, crossing the meadow to the creek and on, out of the valley. She wondered if they were sick, if some disease was going around. I wondered the same thing that afternoon when Philmoore, my springer spaniel, started to shake and whimper. Recalling Julie's story I made a vet appointment for the following morning. An hour later, the chickens, probably the last animals to know before us, began to make an unholy racket; this lasted until 6:52 p.m.

I know the exact time, for that's when the ground rippled and shook; three minutes later my chickens were gone. The geologists would say it was a mild quake, only 4.5 on the Richter scale, but the mountain wouldn't agree. It shuddered—a visible up-and-down trembling that looked to me like a drunk trying to hold a glass steady.

Immediately after came an ear-popping crack, and the entire south side—some 700 million tons of rock and ice—tore free. It came away as if cut by a surgeon's scalpel, a single mass coming down for a night on the town.

The noise it made would have caused a banshee to cower and hide; Roberts, who owns the lumberyard, said later it was a noise so great his ears refused to listen to it. Hidden within the noise was a wild mass of rock and ice that sped down the mountain, the town directly in its path.

From where I stood, across the valley and some 300 feet (91 m) above its broad floor, all I could see was a great, gray shimmering cloud that appeared to settle on the town like a manta ray onto the seafloor. Within this cloud, jumping and whirling like mad dancers, were huge boulders, great chunks of ice, and more. There, before my disbelieving eyes, danced lumber, roofs, cars, and the front of the Royal Vic hotel.

The cloud, having had enough of the town, rolled outward like lake fog. Across the valley it came, boulders cracking and banging to the insane roar of the rushing dust. With the power of a great ship the debris plowed into the valley wall; the trees around me trembled from the impact.

Most of the rock mass turned and ran away, down the valley to the river and on toward the sea. Later the experts said it traveled 60 miles (97 km) downriver and, as it went, sucked up water like a sponge, to become a great flow of mud and rock that destroyed everything in its path.

But that wasn't my problem. Not my worry. My problem was the part of the rock slide that was trying to climb the valley wall. It seemed to inflate and surge upward, coming toward me with a boom and a blast. The blast was wind, pushed before the swirling cloud to knock me flat onto my back. Rolling me over several times, like a dry leaf, the wind skipped on to blow the windows out of the cabin. Ripping the top off the henhouse, it swept the chickens away like they were crumbs on a tablecloth.

Picking my battered body up I ran. No, that's not quite right. With the terror and horror that wrapped itself around me tighter than a hangman's noose, I followed the wind up the hillside faster than I had ever moved before.

As I ran I looked back. A great gray wall obliterated the cabin, the garage, the woodshed, and still was not satisfied. On it came, jetting muddy rocks in front of it like baseballs from a pitching machine. The wall pursued me up the hill until, when no more than a soccer field away, it deflated like a burst balloon. The flow slowed, gurgled, came to a stop, and receded. As it went it left behind a wasteland of broken and uprooted trees, boulders, ice blocks—one with a red skirt hanging from a jagged edge, pieces of lumber, a Ford station wagon, a school bus, and everywhere, thick, clinging mud.

I had lost everything, everything except Philmoore, who panted and trembled at my side. But I was one of the lucky ones—I was alive. Across the way and above me sat my closest neighbor Marley, and he was crying. Going over to comfort him all I could get him to say was "why Lord, why," and he said this over and over. Turned out when he and his wife saw the wall coming, he headed for the backdoor, his wife for the front. She was swept away.

Such stories have been told and retold over the ages, from places as far apart in time and space as Frank, Alberta; Elm, Switzerland; Yungay, Peru; Cartgoa, Costa Rica; Anchorage, Alaska; and hundreds more. All places where great masses of rock, mud, soil, and/or ice have fallen, slid, or flowed. All places where the late Jackie Gleason could truly have cried, "And away we go."

Geologists refer to the "away we go" phenomenon as mass wasting. Mass wasting, to some people, brings images of dieting; in rock circles it brings visions of rock and soil moving down slope under the influence of gravity. Gravity, the common earth hero, does its best to make everything equal. The universal leveler, there is no escaping it; it constantly tugs and pulls on everything, everywhere. Obviously such pulling and tugging is far more noticeable, as well as dangerous, on a surface that slopes compared with one that is as flat as an ironing board. This being the case, people living in sloping states, like Oregon, Washington, Idaho, California, and Wyoming, have the potential to be exposed to more mass wasting then those flatlanders in Iowa, Illinois, and Minnesota.

To demonstrate this all one needs to do is place a marble on a slope and watch it roll away. Put a stone on a steep hill and it wants nothing more out of life then the thrill of sliding all the way to the bottom. Place the same stone in an Iowa cow pasture and it will grow corn before it moves the length of a hog's nose. In a place like Minnesota, however, the stone simply vanishes into one of the nation's 10,000 wetlands.

When you know the hill behind your house is on the move, it is kind of eerie. Looking at a smooth, vegetated hillside, one would never suspect that there is any geologic activity taking place at all. However, if the vegetation is stripped away, what remains is soil that is packed full of small rock particles. These bits and pieces, if one cared enough to look, are derived from rocks that are only found farther up the hillside. Movement of material down the hillside has, and is, taking place. Imperceptible on a-day-by-day basis, but, like the steady tortoise, it is constant and sure. Such movement goes on wherever earth processes, or humans, have carved a sloping surface.

What the earth builds up, the earth, with help from gravity and its two accomplices, water and wind, tears down, in a constant remodeling of the planet's surface. In some places the tearing down is ever so slow; in others, it comes and goes like the rising and falling of the tide; whereas in still others, it is faster than a ride on a giant rollercoaster. Needless to say, when people build and live amongst this constant earth change, disasters will occur, over and over; hills will slide, roads will crumble, and the mountain will fall. It's all in a roll of the dice, so take some earthly advice: Fences won't help, and if the dog suddenly leaves home, don't be far behind.

STIRRED AND SHAKEN

Rocks falling here, soil slipping there, mud sliding all around; if you happen to live on shaky ground you never know who's coming to dinner.

These various dinner guests come with a variety of name tags, but to geologists and engineers, most of them can be called landslides or slumps and creeps. Seeing is believing as far as landslides are concerned, for they are the visual, down-slope movement of rocks and/or soil. Slumps and creeps, by contrast, are the imperceptible down-slope movement of the same earth material. Landslides, along with slumps and creeps, cause more than 3.5 billion dollars' worth of property damage in the United States each year; this means not only do they come to dinner uninvited, they also never pick up the check.

Landslides are quick-moving masses of rock, soil, and sometimes ice that slide, fall, and/or flow down slopes. They give little or no advance warning that they are coming and are the most spectacular and destructive of all mass-wasting processes.

Slumps and creeps involve ground movements that take place over a period of days to hundreds of years. The slow creep of soil down a hill tilts fences, power poles, grave stones, and trees; the quiet, snail-paced sagging of ground produces an escalator-like surface. Just like bad dinner guests, these two are always late, but they are also always on the way.

To find out why landslides slide, why creeps are creepy, and why slumps slouch, we need to go to the original earth cookbook and look up the recipes for slide stew and creep casserole. In doing this it turns out that the basic preparation and the starting ingredients for both are the same. To make slide stew or creep casserole

The slow, downward creep of soil near Park City, Utah, led to the cobra-like shape of these trees. (Photo by R. W. Ojakangas)

the earth starts with a tilted surface—a slope, or incline. The surface doesn't have to be a neck-breaking ski hill either; a gentle slope that a frog could bump up will do just fine.

Once the slope is present and accounted for, next come the ingredients—the right rocks and soils, which are the heart and soul of both dishes. To make slide stew the right rocks are needed; rocks that were made to slip and slide, to roll and bounce—rocks born to wreak havoc. Such rocks start out normal enough, but end up being deeply weathered, badly broken and fractured, or worn round and smooth. These kinds of rocks are much easier to move down slopes than solid, upstanding ones. Smooth, round rocks, for example, can be rolled and bounced down slopes like ball bearings off the hood of a car. These are much preferable to rough, angular rocks that stick to slopes like they were put there with crazy glue. Finally, the right rocks can be ones that are stacked on top of one another like layers in a cake. In layer cakes it is usual to have soft, spreadable layers of rich frosting sandwiched between layers of more solid, crumbly cake. Layered or bedded rocks are similar, with soft, slippery rocks commonly spread between rocks that are hard and solid. The hard layers may then slip or slide over the soft ones given the proper earth conditions.

To make creepy casseroles, the main ingredient is simple and plentiful—dirt. Dirt that is black and full of humus, red and full of clay, or gray and sandy; any or all of these will do just fine.

The final ingredient, water, is entirely optional; it may be added by the spoonful or the bucketful or entirely left out.

As it seeps into rocks, water fills up the small open spaces between individual mineral grains (called pore spaces), reducing the friction between the grains and making the rock more slippery than a snowboard on ice. Water may also have another roll to play. By being present it increases the weight of the rock. This extra poundage may be all that is needed to make the rock mass unstable and send a mountain to the town.

Within unconsolidated material, such as soil and clay, the role of water is more variable. A small amount can actually make the soil more sticky and thus less likely to slip or slide. This is similar to adding water to bread dough to get all the little dough bits to stick together in one kneadable mass. Add too much water and it's back to the flour bin—the dough has turned into a soggy mess. Too much water greatly reduces the friction between dough bits or, on a slope, soil particles, and away we slide.

If the soil covering a slope contains a lot of clay, as is common in glaciated areas, the soil may be able to absorb large quantities of water, much like college students drink beer at a kegger. In fact, there are some clays that can absorb up to 20 times their weight in water. This far outdoes even the most determined partygoer, but not Bubba, the big, old boa constrictor. Bubba can consume an entire pig that is several times his body weight. However, unlike the great snake, which is now so bloated and lazy he won't budge for a month of Sundays, the water-stuffed clay turns into a slippery, slimy gel. It doesn't take much to send this stuff sliding down slope.

Such clay-rich soils are referred to as expansive clays. They could also be called expensive clays for they are one of the most costly geological hazards in the contiguous 48 states. The sliding of expansive clays causes more than 2 billion dollars in property damage each year.

It's difficult to believe, but to help put a halt to the slipping and sliding the heroine called upon was a real live weed, the hero a sharp, thorny bush. Dandelions to the rescue; stinging nettles save the day; indeed stranger than fiction, but many times true. Weeds, thorny bushes, and creeping grasses, in their many splendid coats, cover and protect many a slope. In doing so they often spoil slide stew and creep casserole by providing a strong, interlocking root system that has the ability to hold unconsolidated material (wet soil or broken rock) together, and prevent it from sliding and slipping. The vegetation also slurps up water from the soil and, like flour added to wet dough, reduces the overall moisture content, helping keep dirt in its rightful place.

Temporarily. For look who comes with shovels and bulldozers and chain saws to clear, burn, build, log, beautify, and urbanize. And who pays the price when the cleared, landscaped hill ends up in the swimming pool or on the dinning room table.

The recipe has been read, the preparations made, and the right ingredients chosen and mixed. Now they sit, simmering and stewing for a few days, a few years, a few centuries—but nothing happens. The hill refuses to creep or slouch, and the mountain won't fall because, like Frankenstein's monster, a life-giving spark is needed to start the whole concoction moving.

In the case of slide stew and creep casserole there are four main energizers. The first can be found in the words of the superspy, James Bond, who when ordering his favorite drink always specified" shaken, not stirred." The hill or mountain will take it either way—a good shake or a fast stir to set the mix in motion. And what better, and more ready, to do the job than a bone-rattling earthquake? The stirring and shaking of hills and mountains can force apart, or liquefy, soil particles and break, crack, or pulverize masses of rock.

On a slope, this greatly reduces the friction that holds the material together and, if that happens, it's off to the races we go.

In 1929 a 7.7 magnitude earthquake, in the northwest part of the South Island, New Zealand, started more than 1,800 landslides in a 5,000-square-mile area (12,880 km^2). A similar event occurred in 1971 in the San Gabriel Mountains of California. An earthquake triggered thousands of landslides, which caused the formation of numerous dust clouds that hung over the shaken area for days.

Landslides sparked by earthquakes do not necessarily occur immediately after the mountain has been severely shaken. Instead, what the earthquake may do is bring a mass of rock and soil close to the failure state. The actual rupture and slide can occur weeks or months after the earthquake has gone home. This leaves everyone shaking their heads, wondering how it ever happened.

After earthquakes, rain is the next best thing to get a slide moving. Heavy downpours saturate soils and rocks with water, and this can quickly bring them to the critical stage of instability.

It was days of heavy rain, combined with spring runoff from the snowpack, that triggered a large landslide in the Gros Ventre River basin of Wyoming in 1925. An estimated 50 million cubic yards (38 million m^3) of rock and soil slid into the river, forming a natural dam some 230 feet (70 m) high. Two years later this dam collapsed, causing a flood and several deaths.

Heavy rains may also saturate the soil with water. A subsequent earthquake can then turn the soil and water mixture into jelly and send it happily on its way. This is what happened in the Kansu area of China in 1920. The hills of Kansu are composed of fine, soft wind-bone deposits called loess. The violent shaking of this material caused it to liquify and flow. This led to landslides that swept over numerous villages killing more than 200,000 people. In the summer of 1994 torrential rains in India triggered massive landslides, and widespread flooding that left 617 people dead, and 850,000 homeless.

Not the least exciting energizer is the natural or artificial oversteepening of a slope or hill, which can cause the rocks and soils on it to become critically unstable. Natural oversteepening is

The landslide "scar" formed by the Gros Ventre landslide, Teton National Forest, Wyoming. (Photo by R. W. Ojakangas)

caused by streams, which undercut the banks of slopes, and waves, which pound against the base of cliffs, making the tops fall into the sea. Oversteepening is artificially brought on by the removal of material from the base of a slope during the construction of highways, roads, ski hills, subdivisions, shopping malls, and so on. Removing too much material from the base can take away the support necessary to keep the upper part of the slope in place.

The least exciting energizer of landslides is the washing and drying of soil, which is the major cause of the creeping of soil down a slope. The wetting or freezing of soil particles inflates them like oatmeal in hot water, causing the particles to expand upward at an angle to the slope. When the particles thaw out or dry off they shrink back to their normal size, but not to their normal place on the slope. The thawing or drying, combined with the downward tug of gravity, forces the shrinking particles a fraction of an inch down the slope. This goes on season after season, year after year, decade after decade, until one fine spring morning, the flagpole is parallel to the driveway.

ALL DOWNHILL FROM HERE

Fallers, sliders, and slumpers may sound like the three stooges of mass wasting, but they are about as far removed from comedy as Iceland is from the tropics.

Fallers

Fallers are the bungee jumpers of the mass movers. They leap off tall cliffs, free-falling straight to the bottom; or they bounce down hillsides and mountains, leaping and bounding along like ballet dancers crossing a stage.

Fallers are mostly rocks. They can be brave, single stones, large, bragging boulders, or enormous rock gangs with a mass in excess of a hundred million tons.

Rockfalls are a normal part of mountain life. They are constant pests, blocking highways and roads, covering railroad tracks, destroying bridges, and damming streams. Even if you have never been annoyed or delayed by a bungee jumper, you may still know the hazard they present because of the signs. The ones posted along narrow, curving roads that seem to drop off into infinity on one side and have a sheer rock wall on the other; you know, the ones all those big logging trucks use. The hazard signs show a rather large boulder bouncing down a sloping surface, small pieces breaking off as it goes, and they warn Danger! Caution! Even though I've seen them dozens of times they still are a pain in the neck, for I am forever stretching mine to look up the mountainside to make certain nothing wicked rolls my way. I also found out it is hard to do this and watch for logging trucks at the same time.

Bungee jumpers have the ability to turn into killer landslides or avalanches because of their high momentum. And they may be set into motion by a far-removed earthquake.

The earthquake had a good alibi; indeed it was far from the scene of the crime—81 miles (130 km) from Nevados Huascarán, the highest mountain in the Peruvian Andes. But at magnitude 7.7 it was still strong enough to shake loose a great mass of rock and ice that ended up burying two towns and killing 18,000 people.

Once on the loose, the rocks and ice bungee jumped some 1,900 feet (579 m) to slam into the lower slope of the mountain. The collision pulverized the debris, turning it into an avalanche of budget deficit proportions. Moving at more than 200 miles per hour (322 kmph), the avalanche swept down the mountain and across a glacier that filled the valley at the mountain's base. The speed of the rock mass caused rapid, frictional melting of the ice on the glacier and the incorporation of a large amount of water into the avalanche material. The water increased the mobility of the moving mass, allowing it to run another 8 miles (12.8 km) to the Santa Rita River.

Avalanche meeting river was love at first sight. The two thoroughly mixed, creating a colossal flow of mud and rock that continued down the river all the way to the sea, a distance of 75 miles (120 km). Before meeting the river, the avalanche, all on its own, destroyed a 10-mile-long by 3-mile-wide area (16 × 5 km), burying the towns of Yungay and Ranrahirca. The avalanche was powerful enough to flick 7-ton boulders out its sides like a person flicks away gnats. The boulders, as if launched from a canon, flew up to 2.5 miles (4 km) before impact. Unfortunately, they slammed into houses and buildings, knocking them down like a bowling ball knocks over pins, and into roads, gardens, and hillsides, giving the area the look of a World War II battlefield. The largest crater found measured 108 × 50 × 23 feet deep (33 × 15 × 7 m).

Sliders

Sliders, like bite-size burgers, go down easy and fast. They move as a coherent mass along well-defined surfaces, like kids in inner tubes coming down a water slide. Consisting of rocks, soil, snow, and/or ice, sliders commonly get their start in layer-cake rocks, places where stronger, more competent layers are able to slide over weaker, softer ones.

Sliders range from bite-size burgers, which move only a few cubic yards of material, to double whoppers that involve millions of cubic yards of rock and soil. Bite-size sliders travel at the rate of a few feet per week to feet per day, whereas whoppers can motor along at speeds in excess of 80 miles per hour (129 kmph).

Nobody believed Turtle Mountain would be a slider, let alone a whopper, certainly not the residents of Frank, a small coal-mining town in the southwest corner of Alberta, Canada. The town sat directly in the shadow of Turtle Mountain, which, as it turned out, wasn't a candy mountain after all but a giant layer cake mountain. There were hard, massive layers of limestone (the same stuff coral reefs are made of) sandwiched by soft, weak layers of coal-bearing shale (clay-rich rock).

The fact that Frank mined coal was most likely what pushed the mountain over the edge. The coal mine was located right at the base of the Turtle. To get the coal out large chambers were carved into the Turtle's shell, rooms that were 130 feet long (40 m), 250–400 feet high (76–122 m) and 15–20 feet wide (4.5–6 m), about the width of the coal-bearing layer.

In the summer of 1902, 7 months before the disaster, the miners started to hear strange creaks and spooky groans that seemed to come from the upper part of the rooms; at the same time the mine started to shake ever so slightly. It was soon realized that the noises were caused by the slow slipping of the hard limestone over the softer shale, which caused some of the miners to shake in fear. Taking it for an omen, they packed up and left. Three months later the mine became more automated than a modern car factory, for the miners claimed the "coal was mining itself." The soft, black coal had started to bulge out like a bad bicycle tire. The movement of the limestone was squeezing it out of the shale layers like toothpaste from a tube.

The strange music, the slipping, and the automated coal mining continued until 4:03 a.m. on April 23, 1903—that's when the tire burst. The entire east side of the mountain, some 40 million cubic yards (30.5 million m^3) of hard limestone, slid rapidly over weaker shale and fell into the void. The rocks leaped, slid, bounced, and pulverized their way down 3,000 feet (914 m) of mountain at more than 60 miles per hour (97 kmph). Reaching the bottom, the debris tore across a 2-mile-wide valley and climbed more than 400 feet (122 m) up the opposite valley wall. In its rush to cross the valley the slide raced over the outskirts of Frank, killing 70 people, destroying numerous houses and cabins, and burying more than 7,000 feet (2,134 m) of railroad track. The entire event lasted about 100 seconds.[1] The deposit left by the slide covered the valley and was typical of the mess left by sliders. A wild jumble of various-sized rocks, many more than 10 feet across (3 m), sat in a gooey mass of pulverized rock dust, mud, trees, soil, and railroad track; it looked a lot like rocky road ice cream.

Though the slide killed 70 people, there were many more who escaped by the seats of their long johns. One of these was a Mr. Warington, who was sleeping in a cottage on the wrong side of town. Early in the morning he was awakened by a loud noise, which he believed to be hail pelting down on the roof of the cabin.

Leaping out of bed for a looksee, he immediately realized it was a lot more serious than ice from the sky. It was rocks from the mountain, and as this was sinking in, the cabin started to rock back and forth and to make fearful groaning sounds like "a hungry bear coming out of hibernation." The next thing Mr. Warington remembered was standing up smack in the middle of a great pile of boulders. He was some 40 feet (12 m) from where the cabin had stood, his bed was 20 feet (6 m) farther on, and the cabin had vanished. His leg was broken and he was covered with bruises and scrapes, but he was alive with no idea of how he got to where he stood, or how he woke up safe and sound in the middle of a boulder field.[2]

It is interesting to note that the width of the Frank slide was about the same as the width of the coal mine. It is probable that the excavation of the caverns, to extract the coal, removed enough support from the base of the mountain to allow the top to go the whole nine yards.

Sliders that float on air, like a genie on a magic carpet, are not part of a Disney movie, but rather avalanches of rock and ice that start out by free-falling down steep mountain slopes. As the falling mass nears the ground, air that has been pushed ahead of it gets trapped and becomes compressed beneath it. On impact the trapped air forms a thin cushion that allows the rock debris to move buoyantly over the ground surface, much like a Hovercraft moves over water. Because of this deluxe cushioned ride the avalanche has the potential to accelerate to speeds greater than 125 miles per hour (201 kmph) and to end up tens of miles from the mountain.

Fortunately, avalanches that can levitate are as rare as a January heatwave in Minnesota. Nonetheless, when they do occur, they are armed and dangerous, for they are fast and typically generate the movement of millions of cubic yards of rock debris.

In 1917, as World War I was grinding to a close, a large mass of rock and ice fell from the crest of Mt. Blanc, located on the Italy–France border and the subject of a Percy Shelley poem. The mass fell onto the Troilet Glacier where it pulverized on impact. The smashed and disintegrating debris surged down the glacier, riding on a trapped cushion of air. It raced into the valley below the glacier's snout, destroying two towns and obliterating several cabins over 200 feet (61 m) up the far valley wall. The entire event lasted 3 minutes. More material was moved over a longer distance in that short time than the two warring armies moved in 2 years of trench fighting.

Halfway around the world, and 47 years later, the great Alaskan earthquake shook the ground so hard that 87 miles (140 km) away a tall mountain peak, now named Shatter Mountain, did just that. A huge slab of rock, some 2,000 by 3,000 feet in size (610 × 944 m), broke away from the mountaintop and bungee jumped more than 1,600 feet (488 m). Slamming into a lower slope, it disintegrated into a large avalanche of boulders, dust, and ice.

The valley at the foot of the mountain was broad and U-shaped due to glacial erosion; its occupant was the Sherman Glacier. As the avalanche tore down the mountain toward the glacier, it first had to negotiate a high ridge that sat between the glacier and the mountain. The top of the ridge was 82 feet (25 m) above and directly overlooked the glacial ice. Upon impact the rock debris was launched, en masse, straight over the top of the ridge, clearing it as easily as a motorcycle stunt driver flies over cars. This leap created a flying, sheetlike mass of debris that, when it finally landed on the glacier, had a broad pocket of air trapped beneath it. The launching of the debris into orbit, along with the air-cushioned ride, allowed the avalanche to cruise along at more than 175 miles per hour (282 kmph). Across the glacier it flashed, and on for some 20 miles (32 km), carrying more than 30 million cubic yards (23 million m^3) of material. Fortunately, this event occurred in a remote area and no lives were lost. Think what a similar slider could do to a populated area. Skiing anyone?

Mount Rainier, a broad, Cascade teapot volcano, is certainly closer to home, but it is still relatively far removed from what can be called urban civilization. Good thing, too, for in 1963 a large part of Little Tahoma Peak fell 1,700 feet (518 m) down the mountain to collide with the Emmons Glacier. Buoyed by a cushion of air, the avalanche material shot across the glacier at speeds in excess of 100 miles per hour (161 kmph). At the end, or snout, of the glacier the entire mass of rock and ice soared into space. More than 14 million cubic yards of debris soared over a 6-foot-high (1.8 m) stream gauge station, leaving it completely untouched! Now that would have been a picture worth more than a million words. The avalanche finally came to rest about one-half mile (0.8 km) from the White River campsite—some 5 miles (8 km) from the source of the landslide. In traveling those 5 miles (8 km) the rock mass dropped some 6,200 feet (1,890 m) in elevation, carrying boulders measuring as large as 60 × 130 × 160 feet (18 × 40 × 49 m) and weighing up to 50,000 tons.[3] Imagine something like that paying a surprise visit to a large city nestled in the shadows of a beautiful, serene mountain?

What on earth could possibly top avalanches that float on air? How about sliders that catch on fire, like the burning and smoldering masses of shale that slide down steep hillsides on the west coast of Greenland. Sliders afire were first reported by Viking sailors more than 700 years ago. Sailing up the Greenland coast, poking in and out of the many fjords, the Norsemen saw columns of smoke rising high into the sky. Coming closer to land, they saw flames mixed with the smoke, and it all seemed to come from hillsides of mostly bare rock. The vikings named the place Eysunes—smoldering, rocky promontory. Since Viking days the shale has continued to slip, slide, and burn. Strange as it seems there is a direct connection between sliders and burners, for shale that stays in place never catches on fire. It is believed, though not proven, that the disruption and sliding of the shale allows air and moisture access to the bituminous (coal) layers in the shale, and that the heat generated by the friction of sliding is enough to cause spontaneous combustion—great hills of fire.

Slumpers

Then we have the slumpers. No, I'm not referring to teenagers sitting on couch. The slumpers I'm referring to belong to the earth. A slump is the slow (feet per day), downward slip of a coherent body (a mass of rock or soil), creating a surface that ends up looking

Watch your step. The downward slippage of a coherent body of soil to create a surface that looks like a broken escalator, slumps typically have clifflike scars at their tops. (Photo courtesy of R. W. Ojakangas)

like a broken escalator. Slumps typically have a clifflike scar at the very top.

Slumps are a common form of land movement on slopes that have become oversteepened. The rocks and/or soils at the top of the slope are held in place by the material at the bottom. If enough material at the base is removed, either by streams, wind, waves, or people, the top will become unstable and sag like a teenager whose CD player just quit on her.

STONE SOUP

Do you know the children's story *Stone Soup* by Ann McGovern?[4] The tale is of a tired, hungry wanderer who stops at what he considers to be a fine house to ask for food. The woman who owns the house replies that she has nothing whatsoever to give him. He then asks if she has a stone so he can make stone soup. Once he has obtained a stone he then asks for a pot full of water so he can cook his stone soup. Placing the stone in the pot he sits back and waits for it to boil, bubble, and cook. By the end of the story the wanderer has managed to get the woman to put carrots, onions, bones, and barley into his stone soup to make it a truly delicious stew fit for a king. Each time she adds another ingredient she says, "soup from a stone, fancy that." In the end the wanderer goes on his way, thinking about his next night's meal and, as he does so, he repeats the woman's words, "soup from a stone, fancy that."

Even though this wanderer managed to talk his way into vegetables and bones for his soup, the two main ingredients were water and a stone. The earthly version of stone soup is similar; water and stones are the main ingredients, but, like the hungry wanderer, the earth manages to get lots of other goodies thrown in. Geologists call stone soup made by the earth a mudflow.

Earth-made stone soups are relatively fast-moving slurries of water and rocks with all sorts of other debris mixed in: soil, trees, ice, sheep, asphalt, power poles, skiers, cars, and trains. Their consistencies range from wet concrete, sort of like a Minnesota hot dish—lots of filler but little sauce—all the way to chicken broth. The rate a mudflow will travel, as well as the distance to its final destination, depends on the steepness of the slope it flows over, and the amount of water it contains. In general the steeper the slope, and the higher the water content, the faster and farther the mudflow will go. Mudflow speeds range from choking gurgles, which move a few feet per hour, to mad, roaring rushes of more than 100 miles per hour (161 kmph). Distances traveled are anywhere from a small city block to the width of a moderate-size state (150 miles [242 km]).

Mudflows are particularly dangerous hazards in two geologic situations. The first is in dry, mountainous areas. Here heavy rains can wash rocks and soil off lightly vegetated slopes into gullies and canyons, creating a surging wall of water and mud that cascades down the canyon picking up, or knocking over, everything in its

Rocky deposit left by a stone soup flow that contained abundant boulders and trees but little water.

path. Upon exiting the canyon the watery wave quickly deflates, spreading out to form a wide mudflat that can be anywhere from 1 foot (0.3 m) to more than 100 feet thick (30 m).

The other favorite place for mudflows to hang out are on the slopes of volcanoes; any volcano will do, be it an old turtle or a young teapot. Mudflows that originate on volcanoes are given the special name *lahar*, which is Javanese for volcanic mudflow.

Stone soups from volcanoes are terrific hazards, for they can occur during volcanic eruptions, as well as long after the volcano has apparently become dormant. There have been more than 100 stone soup flows off of Mt. Rainer since the end of the last ice age, about 10,000 years ago. The so-called ring plains, pancake-like flats that occur around the base of many a teapot volcano, are dominantly composed of stone soup deposits.

Stone soup that flows down and off volcanoes can get started in a variety of ways.

Take an old turtle volcano and erupt lava out of it so fast that much of the lava piles up along or around the cracklike opening until the volcano becomes steep and heavy. Then, with an earthquake here, a heavy rain there, the steep, heavy part of the turtle slides away into the sea. The remains of such mudflows have been recognized on the seafloor off the Hawaiian Islands. These slide deposits can exceed 77 miles (125 km) in length, and some of the pieces of lava are bigger than Yankee Stadium. Even without George Steinbrenner, this would make a pretty big splash.

Big bang eruptions, through small lakes that may occupy the spouts of teapot volcanoes, can create stone soup flows. Such an explosive eruption causes large amounts of ash and pumice to mix with gallon upon gallon of lake water. The whole soupy mess is blown skyward and, when it rains back to earth, a stone soup flow is born.

This is what happened when Kelut volcano in Java blew its top in 1919. Forty nine million cubic yards (38 million m^3) of water were blown out of a lake that had formed in the teapot's spout. Along with this came ash, pumice, and rock fragments courtesy of the volcano. The whole soggy mess flowed down the slope of the volcano to bury 51 square miles (131 km^2) of farmland, wipe out many villages, and kill over 5,000 people. The mudflow traveled more than 25 miles (40 km) from the volcano.

On the north island of New Zealand sits an active teapot called Ruapehu. Ruapehu has a small lake at its summit that bubbles and boils from the heat of near-surface magma. Periodic eruptions through the lake create stone soup flows that radiate down and away from the volcano. Strange as it may seem, New Zealanders have built a ski resort on the south side of the volcano! Mudflows of years gone by have, on more than one occasion, wiped out the ski lifts, forcing the owners to move them to a safer location. To give skiers some reassurance, an eruption alarm system has been installed. When Ruapehu erupts a bell rings, telling skiers they have about 8 minutes to get off the slopes. If they fail to do so, they may well become one of the extra seasonings thrown into the wandering mudflow. This certainly adds a new wrinkle to the sport; it might also make for a new, and very exciting, Winter Olympics event.

Stone soup flows can also form when snow ice melts on the top and flanks of a volcano. This can happen when eruptions take place directly through ice and snow, when hot, glowing avalanches (pyroclastic flows) travel over snow or ice, or when magma rises high enough toward the surface to cause large-scale melting.

One of the greatest mudflows in postglacial North America formed when a glowing avalanche rushed down the side of Mt. Rainier some 4,800 years ago. The pyroclastic flow melted ice and snow on the volcano's flanks, and then it incorporated a lot of the meltwater, to be born again as a mudflow. The mudflow raced on for some 65 miles (105 km), traveling all the way to Puget Sound. Here it ran out of steam, deflated, and spread over the lowlands to form a deposit of mud and rock some 19 miles long (30 km) and between 3 and 11 miles wide (5–17 km). Some of these stone soup lowlands have now been developed as part of the greater Tacoma area in the state of Washington.

In 1985, the Colombian volcano Nevado Del Ruiz exploded, forming glowing avalanches that melted the ice and snow that capped the mountain. A torrent of viscous mud, ash, and rock radiated away from the summit area as hot mudflows and traveled down the volcano along existing stream channels that were filled to the brim with rainwater. One mudflow moved rapidly down the Laqunilla River and devastated the town of Armero, some 30 miles (48 km) from the volcano. There was no advance warning, and 20,000 people perished.

There was an advanced warning, however, on Christmas Eve 1953 in New Zealand.

The lone figure stood beside the railroad tracks. Lantern in hand, he waited. The collar of his long coat was pulled up around his ears, his body bent, shielding the flickering light from the wind. The noise was fearful. The roar of the water and the banging and clatter of the rocks made him think of hell and redemption. Saints preserve us, he thought, hell it is, and on Christmas Eve. He had to stop them, had to . . .

But he couldn't hear. The noise was too great. And before he knew it, the single beam of bright light swept over him like a prison searchlight. The train sped around the curve.

Raising his lantern he frantically waved it back and forth, up and down. "Stop! Stop!" he yelled. "In the name of God, you must stop. There's no . . . "

The engine vanished into the noise and the night. The cars rolled past him, faces lit against the glass. A young girl, red cheeked, laughing; a man wearing a tall hat, drinking from a crystal glass; a boy, sucking licorice, sticking out his black tongue at a small girl. Gone, gone, into the blackness.

Throwing his lantern to the ground, he screamed and fell to his knees, crying. Out of the night came the sounds of metal and water and rocks and screams—oh God, he'd failed, but what else could he have done?

On Christmas Eve, 1953, the Wellington–Auckland express train raced on to the bridge at Tangiwi and plunged into the waters of the Whangaehu River. A volcanic eruption 20 minutes earlier had caused the sudden discharge of thousands of gallons of water from the crater lake at the top of Mt. Ruapehu. The swift and sudden deluge of ash, boulders, and debris, a stone soup flow, hurled itself against the concrete pylons of the railway bridge and swept away 154 feet (46 m) of track. As the train raced onto the broken bridge, the engine and fuel car smashed their couplings and crashed into the opposite bank, 118 feet (36 m) away. The passenger cars followed and fell 35 feet (10 m) into the river. Six cars were quickly submerged and swept downstream. Out of 285 people known to have been aboard the train, only 134 survived. A total of 131 bodies were recovered and 20 others were never accounted for. Survivors spoke of a lone person who attempted to flag down the train with a lantern, but to no avail.

The last way to make stone soup at a volcano is to cover or bury the natural vegetation on its flanks with thick deposits of ash. Without the vegetation to slow runoff, heavy rains can quickly turn the ash into mud, making a fine stone soup. This instant mud pie then flows downslope to wreak havoc on anything that gets in its

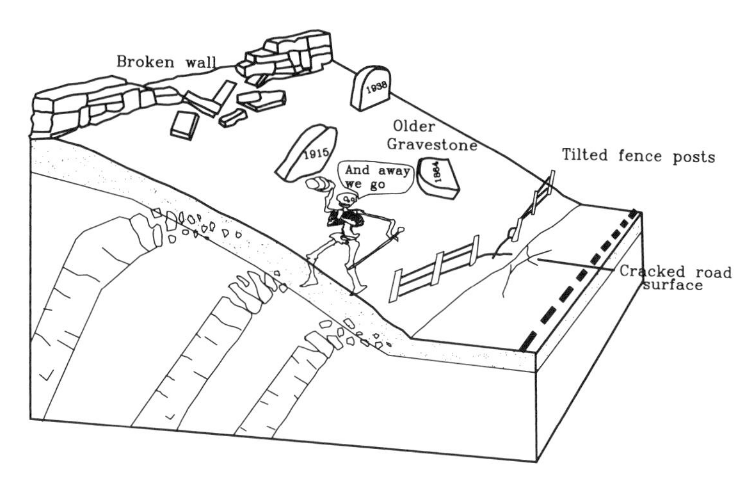

Slow but creepy. The imperceptible, continuous downhill movement of soil and/or rocks is called creep. The only indication that anything is happening is the tilting of old gravestones, the cracking of roads, the breaking of fences and walls, and the tilting of power and telephone poles. (Adapted from F. Lutgens and E. Tarbuck, Essentials of Geology, 1992)

way. In 1963, the city of Cartago in Costa Rica got in the way of one of these slurries and was destroyed.

EASY DOES IT

The gradual, mostly invisible, downhill movement of soil is called creep. However, something taking place you can't see should at least be called creepy if not downright scary. Here you are, in your new house, a beautifully landscaped, grassy hill rising up right behind your backyard. Then one day you notice the fence has started to lean, the power pole is tilting and the driveway has cracks all across it. Only then do you realize that something wicked this way creeps; the hill is ever so slowly moving into your backyard.

Creep hangs loose on most soil-covered hillsides. The rate of creep is higher on steep slopes and in wet soils than it is on gentle

The downward movement of soil caused by the melting of ice in the upper few feet of the soil. Since the soil below this remains frozen, the water is forced to run downhill and carries the soil with it. This leads to a hillside with the appearance of an old, rumpled carpet. (Photo courtesy of C. L. Matsch)

slopes and in dry soils. However, a thick cover of vegetation will slow or greatly impede creep because plant roots do a great job of binding broken rocks and soil particles together.

A process closely related to creep is solifluction (no, this is not a religious term). Solifluction occurs in areas of high latitude where large portions of the soil remain frozen year round. During the spring and summer months the ice in the upper few feet of the soil melts. The ground below this, however, remains frozen and solid as concrete. The water in the thawed part of the soil has only one way to go. It slowly seeps or percolates downslope. Unfortunately, where the water goes so goes the soil. It is slowly carried along with the moving water to give the hillside the appearance of an old, rumpled carpet.

DAM-NATION

People are experts at building and developing on, or right below, landslide-prone areas. They are also very good at giving landslides a helping hand, by clearing vegetation off slopes, logging hillsides, building dams and highways, creating subdivisions, and, on occasion, by strip mining. All of these activities have been going on for so long that it is too late to reduce landslide hazards in populated areas by avoiding them. This being the case, is there anything to be done but worry and pray? In mountainous areas, where slide zones are narrow, structures can be built over roads and bridges to protect them. However, this solution is way too expensive to practice on a large scale (imagine a canopy over the city of your choice). All that can be done in cities is to try to reduce the danger and the destruction *when, not if*, the slide comes. There are several techniques for doing this, the most common of which are the following:

1. Reduce the weight and the angle of a dangerous slope. This is done by removing material from the top and sides, not the bottom. This works well on hillsides, but on a massive mountain—who's gonna do that?
2. Add supports to the base of a slope. This is typically done by using giant rock bolts, which are driven down through the soil and broken rocks into hard, solid rock below. The bolts can then be connected by steel cables to give them added strength.
3. Plant weeds, bushes, and trees and then plant some more. Then keep right on planting, anything and everything that has a deep root system and grows quickly. Some people may complain of the ugly hillside covered with poison ivy, bitter vetch, and stinkweed but, believe me, it makes it a lot safer.
4. Take steps to reduce the water content of soils and rocks by:
   (a) Covering the surface with a substance that is impermeable. It is not uncommon for steep roadside cuts to be covered with concrete or asphalt; however, I don't think this would be very practical for an entire mountain.

(b) Divert the runoff of water by installing drainage pipes; this is relatively easy and effective.
(c) Pump old water wells to lower the level of the water table, thus drying out the surface soils.

Unfortunately, it is terribly expensive to do a lot of these things, and no one wants to pay for protections against something that *might* take place. This puts anyone who lives in landslide-prone areas in a no-win situation. Landslides are part of this dynamic planet and, like volcanoes and earthquakes, they operate on a vastly different time scale than we do. What we, as builders, fail to comprehend, is that what is stable today, next week, or a thousand weeks from now will eventually succumb to a landslide, an avalanche, or a slump. Today, tomorrow, or next season, the piano and trombones will play. Shaken and stirred, that's the way we'll go.

Prelude to Chapter 5

Soft pops and fizzes from bubbling hot springs, whispering gurgles and wheezing croaks from the mud ponds, the occasional roar as a geyser spouts, and around all this, like surround sound, comes the constant hiss and fizz of escaping steam.

If you have the opportunity, take a walk through one of the great hot spring areas of the world, such as Yellowstone, the Geysers area of California, and Waimangu, Waiotapu, and Wairakei in New Zealand. Listen as you walk; listen for the music of the hot springs. Hear the repeated note patterns in clarinets and flutes as the springs bubble and pop, and the flute glissandi as steam hisses from the ground. These repeated sounds are occasionally mixed with a trumpet fanfare as a geyser erupts, bursting skyward. As you walk and listen also keep an eye out for the Greek god Pan. You might just catch a glimpse of him dancing and whirling around the steaming pools as he accompanies the clarinets and flutes on his panpipes.

CHAPTER 5

When the Waters Dance

One by one the fountains spout, like great whales breathing and blowing after a deep, dark dive. All around the splashing waters the wind chases ribbons and spidery clouds of sulfurous steam that rise from bubbling pools and ooze out of the baked earth. Around the geysers are cisternlike holes filled with shimmering green and turquoise waters, pits of black, spitting mud, and deep pools bubbling like champagne in a wide-brimmed glass.

Hot springs, geysers, and mud pots are described as everything from "gardens of hell," "devils gateposts," and "Dante's Inferno" to "fountains of youth," "gifts of the Gods," and "brimstone bakeries." These are the sons and daughters of volcanoes, but are they also like Jekyll and Hyde? Are they cousins of the devil? Nieces to the gods? The waters will never tell: they just dance and the clarinets and flutes play on.

VULCAN'S WATER

Grandma's oven, or Vulcan's forge (the magma chamber to a volcano), is not terribly energy efficient. In fact, it loses heat by the train load and would easily qualify for a home energy loan. This heat radiates away from the oven to be absorbed by waters and rocks of the continental and oceanic crust.

This heated water starts out life as good old rain, river, lake, or seawater. But somewhere along life's journey it runs into a crooked fracture, an evil crack. This meeting sends the water on a slow, downward journey toward grandma's oven. As the water seeps down through the crust, it absorbs some of the oven's radiant heat and begins to sweat and expand. The deeper it goes, the closer to the oven it gets, and the hotter and lighter it becomes. A point is finally reached when the water becomes so hot and expanded it starts to act like a balloon filled with helium, and up toward the surface it rises. Up it comes but, like Jekyll turned into Hyde, the water is now unrecognizable. No longer cold, anemic, and pure, it is now hot, colorful, vibrant, and enriched, for

> Not only water flows from earthly springs,
> But mineral fluids, holding sulphur, iron,
> Acids and gases, lime, and many salts.
> Naphtha and oils from fountains seldom flow;
> Yet there are such, even liquid pitch
> In bubbles bursting underground, in lakes
> Expanding; thro' volcanic regions, prone
> To offer fiery springs, in heat evolving.
> —C. S. Rafinesque

Mr. Hyde's water exits at the surface as hot springs, geysers, and mud pots found in places like Iceland, Japan, Indonesia, Italy, the Philippines, New Zealand, Mexico, and the United States at Yellowstone National Park, the Geysers area of California, and Steamboat Springs, Colorado. Most of these places we have already visited, for they are also the lands where the crust is ruled by volcanoes and earthquakes run free.

It was A.D. 870 when the long ship rowed into the broad harbor. Clouds of thick steam hung over the cold water, so the Viking explorers named the place Reykjavik, or smoking bay. Since the arrival of these first Vikings, the people of Iceland have bathed, washed clothes, and cooked in the many hot springs and thermal pools scattered across the countryside. They heat their houses with the warm water and bake bread in the steaming ground; a long winter's night or an all-day summer, the hot waters always seem to provide.

In fact, as far as smog, fog, and grime go, Reykjavik was right up there with Sherlock Holmes's London. However, in the 1930s, Icelanders started to drill into the steaming ground and tap the hot, subterranean water. At 140° F (60° C) the water is hot enough to be pumped directly into hot-water radiators and tanks. This water now heats about 99% of the homes and stores in Reykjavik—a long, clean cry from the coal- and wood-burning days. The water is also used to heat swimming pools and hothouses year around. It is cheap, efficient, and abundant; Reykjavik was the first city of the world to be heated by planet Earth.[1]

In A.D. 1300 a Maori canoe came ashore on the North Island of New Zealand. Stepping into the cold air the high priest saw and felt snow for the first time. Convinced the land was under a terrible spell he prayed to the gods for warmth. Lo and behold hot springs and geysers appeared to melt the snow and chase away the evil cold. The gods had surely smiled on the Maoris by giving them this great gift, and they would treasure it forever. Since that time the Maori people have used the warm and hot waters for heating, cooking, and bathing. Great Maori feasts, called *hangi*, are centered around "this gift from the gods." Meat and various kinds of vegetables are wrapped in palm leaves and placed in baskets to be lowered into steam holes to cook (a dash of sulfur in the volcanic steam makes the meat extra good). And, much like North American children, who roast marshmallows in campfires, Maori children dunk ears of corn into bubbling pools of hot water, cooking them in seconds.

Animals, as well as people, benefit from hot springs. In Yellowstone National Park, the land of 10,000 hot springs, buffalo, elk, deer, and any other creature that's up on a cold winter's day hang out at the hottest spot in town. Yellowstone (so named because sulfur precipitates out of the waters and colors the volcanic rocks yellow) has been a great tourist attraction since 1872. It was also a popular spot to "come for the cure" or to "take the waters" until the park realized the cure might be worse than the ailment. Now, because of the potentially dangerous elements that naturally occur in many of the hot springs, bathing in them is severely restricted.

Hot springs may also be "farmed" for minerals and power. At Larderello in Tuscany, Italy, the ancient Etruscans mined boric acid from hot springs. The boric acid precipitated naturally out of the cooling steam and was used to make enamels for decorating pottery and tombs (the Etruscans were almost as good at tomb building as the Egyptians). Starting in 1810 the boric acid was commercially extracted. With the advent of commercial mining came the idea of using the steam from the hot springs to power turbines to produce electricity for the farming operation. The first hot spring electrical power was produced in 1904, but the acid gases in the steam attacked and destroyed the metal parts of the steam engines. It was another 45 years before alloys were developed that could stand up to the acid attacks. Since the 1950s Larderello has been capable of producing over 3 billion megawatts of electricity per year. Today at Larderello, ammonia and helium, as well as boric acid, are farmed.

The Geysers hot spring area, some 75 miles (121 km) north of San Francisco, produces more than 1,500 megawatts of electricity from the earth's "sweat." These hot springs were discovered in 1847 by William Bell Elliot, an explorer and surveyor. He was out hunting grizzly bears when he came across something much more fierce and terrifying—a place where steam roared and hissed out of the ground, a narrow canyon only a quarter of a mile long, which Elliot believed was the gateway to hell. By 1860 the Geysers area was a tourist attraction, and 100 years later electricity was being produced there.

FRYING PANS AND HOT TUBS

"I stepped into the warm, bubbling waters. Settling myself it was as though I had returned to the womb. Closing my eyes I drifted along on a soft, sparkling river. It was then I became certain

I was closer to God than any time since I had been born" (comment by a visitor to the mineral spas at Bath, England).

Hot springs are defined as springs of water that issue onto the earth's surface at temperatures "appreciably" above the average temperature of the air at their exit point. This means that if you sat in one in Hawaii you might be boiled like a lobster, whereas in Minnesota you could be turned into a giant ice cube. Appreciably above—now that's real scientific and exact. Some references define "appreciably above" as 11–16° F (6–8° C), others as 15–20° F (8–11° C); the *Dictionary of Geological Terms* says the temperature has to be higher than that of a human body (they don't specify whether that's dead or alive). Now if we grant a warm body, "appreciably" is more than 98° F (36.6° C), which is "appreciably" greater than the average air temperature almost anywhere on earth! Now, as silly as all this seems, this is what helps make geology fun.

Lord Kelvin, hot under the collar one day, said that "geology is about as intellectually stimulating as stamp collecting."[2] Just the fact that earth scientists are able to define the very same term differently, leading to much confusion, some heated arguments, and many government committees, makes Kelvin dead wrong.

So, like typical geologists, we will define hot spring as it best suits us. Here goes. Hot springs are waters that issue onto the earth's surface at temperatures appreciably warmer than leftover pizza. These waters can flow, bubble, seep, or ooze. They can smell as sweet as freshly mown hay, or they can reek like 10-year-old eggs mixed with a wino's breath and served up with oodles of old, cooked cabbage.

The waters of springs warmer than leftover pizza typically contain a lot more dissolved minerals than plain river, lake, or seawater. This is because hot water is a much more effective dissolver of rocks and minerals than cold water. So springs can contain a lot more stuff than a pizza with the works—anything and everything from silica and carbon dioxide to iron, sulfur, magnesium, arsenic, mercury, gold, silver, antimony, bismuth, chlorine, sodium, and much more. These elements can precipitate out of the waters as native elements (such as pure gold, silver, or sulfur), or, more commonly, they combine with another element to form minerals (mercury plus sulfur gives us cinnabar). This precipitation can occur below the surface to sometimes give us a mineral deposit (gold or silver deposits are mined from "fossil" hot springs around the world). Minerals can also precipitate at the surface, to give us boiling paint pots, or continue into the air, to give us pollution.

ARSENIC AND OLD FAITHFUL

Hot springs the world over are of two general kinds: those that make you feel good and those that will kill you.

Hot springs you can lay back and relax in can be called natural hot tubs because they bubble and froth like Schwepps on ice. The bubbles are carbon dioxide (CO_2), and these hot springs occur where volcanoes sit on top of large amounts of carbonate rock (i.e., limestone and/or dolomite); for this reason they are also called carbonate hot springs. If the CO_2 precipitates out of the waters (as a carbonate mineral such as calcite), it forms deposits called travertine or tufa. Travertine can coat objects within the hot spring, such as twigs, rocks, or leaves. In France it is common practice to put religious medallions in the natural hot tubs so they become thickly coated with white calcite; this supposedly makes them more "pure."

If bubbling hot tubs overflow their banks they can build mounds, steep terraces, or tall spires of travertine that can be extensive and quite beautiful. In New Zealand, near Lake Tarawera, such hot springs built the "eighth wonder of the world," the very beautiful and elegant pink, pale blue, and white terraces. Tourists came from all over Europe to bathe in these warm waters, to boat on them, or simply to walk around and admire them. In the spring of 1886 Maoris living around Lake Tarawera reported seeing the phantom Maori war canoe on the lake (the native peoples knew this meant some great disaster was coming). The Europeans pooh-poohed all this as superstitious nonsense. One week later Mt.

Mammoth hot springs, Yellowstone National Park. Travertine deposits in mounds and terraces.

Tarawera erupted, blowing the eighth wonder of the world to smithereens (and some European naysayers along with it).

Natural saunas and steam rooms are hot springs you can not only relax in but also sweat, wash, and even cook in, though not all at the same time I hope. These types of hot springs are called alkali chloride hot springs, for they contain abundant sodium, potassium, and chlorine. At health spas they are said to contain many different "salts," which are believed to be good for bones, skin, and mental fatigue. Unfortunately, some of these hot springs also contain high amounts of radon. The radon is probably derived from volcanic gas that is released directly into the spring. These radon hot springs were once, and in some places (such as the island of Ischia near Naples, Italy) still are, the most popular health resorts in the world. In this case the cure is probably worse than the disease, for radon, a radioactive element, can cause lung and other types of cancer.

Boiling mud pot in New Zealand, called the frog pond because the escaping gas causes the pond to sound like a swamp full of croaking frogs, at Whakarewarewa Thermal Area.

Then there are the hot springs from hell. Drink their water and Maalox or plop-plop, fizz-fizz won't help you one little bit. These types of hot springs are also called acid springs because they are rich in hydrochloric and sulfuric acids, as well as hydrogen sulfide, with or without various amounts of boron, ammonia, mercury, arsenic, antimony, and bismuth, which take paint off of pencils and skin from bones.

Acid hot springs are hotter and contain a lot more of the ingredients that come directly from grandma's oven (volcanic gases and elements) than the other two types. These delightful waters form boiling mud pots, steaming vents, and bubbling pools, the models Dante used for his famous *Inferno*.

These natural paint strippers are also great destroyers of rocks. They attack rocks like bleach does stains and, over time, change

The Prince of Wales geyser, Whakarewarewa Thermal Area, New Zealand.

them into soft, powdery clay or wet, sticky mud. The bottoms of many of these hot springs are made up of viscous mud, which contains lots of iron, sulfur, and other goodies that produce distinctive browns, reds, oranges, greens, and blues; such pools have been called paint pots. If they bubble as well as stink, they are called boiling mud pots; springs such as these are found at Bumpass Hell in Yellowstone.

In some geothermal areas acid hot springs occur right next door to hot tubs and steam rooms. So if you are cooking, bathing, or washing in these waters, it may be simply trial and error as to which kills and which heals.

Geysers are hot springs that simply can't contain themselves. Natural show-offs, they try to outdo each other by throwing columns of hissing water hundreds of feet into the air. The term *geyser* (which means to gush) comes from the great geyser in Iceland. During the 1770s and 1780s, it gushed water 200 feet (322 m) high every half hour. The term is now used worldwide for spouting hot springs regardless of how high the water goes.

Some geysers simply burp water, and others blast it hundreds of feet into the sky. The greatest geyser of them all was New Zealand's Waimangu ("bird water"). Bird water had only a five-year engagement (needed a better agent I guess), from 1899 to 1904. During this time it threw great columns of steam, muddy water, and rocks to heights of more than 1,500 feet (457 m).

The time between geyser eruptions varies from a few minutes to many months and tends to increase with age—a change of life from geyser to geezer to wheezer. Great geyser in Iceland went from 30-minute shows in 1772 to a show every 20 days in 1883. And Yellowstone's Old Faithful just isn't anymore. It used to erupt, on average, every 65 minutes, though the actual time between eruptions was anywhere from 30 to 90 minutes. It is now much more irregular, and some scientists believe it's becoming plugged and will one day spout no more.

In both geologic and human terms geysers are short-lived affairs lasting from a couple to just tens of years. They expire rather quickly due to (1) changes in the water levels of the hot spring system, (2) earthquakes that alter the subterranean plumbing system, (3) volcanic eruptions that destroy geysers and change the plumbing systems, or (4) minerals in the waters that precipitate out to plug the geyser's blow hole.

Today the principle geyser areas of the world are in New Zealand, Yellowstone National Park, and Iceland. Individual or small geysers can be found in most areas where hot springs occur.

THE GOOD, THE BAD, AND THE GASSY

Volcanoes have gas, a huge amount of bad gas that makes them rumble and burp, gurgle and belch, and sometimes just swell until they blow up. Gas before an eruption, gas during an eruption, gas after an eruption—gas, gas, gas. In many instances the most violent, explosive, gas-rich eruptions occur after a volcano has been dormant for a long period of time. The volcanic gas rises upward to collect at the top of the magma chamber. Here it pushes and swells until the volcano can't stand it anymore and blows its top.

Openings on and around volcanoes, from which volcanic gas escapes to the surface, are called fumaroles or, if the gas is full of sulfur, solfataras. Fumaroles can exist and be active for thousands of years after a volcano has apparently become extinct (emphasis on apparently). Fumaroles can be found just about anywhere on a volcano: in blasted out craters, lined up like ducks along snakelike fractures, or scattered willy-nilly up and down the flanks—anywhere gas can easily reach the surface.

Volcanic gas can be a real pain. It can kill crops, animals, and people and can cool the earth to give us cold summers and colder winters. The sulfur in it creates acid rain, the chlorine and fluorine in it help deplete the ozone layer, and, most of the time, it simply stinks. The gas emitted by volcanoes is mostly (75%) water vapor (water from fumaroles on Mt. Erebus in Antarctica freezes around the vent to form towers of ice over 60 feet high), followed by carbon dioxide (CO_2) and sulphur dioxide (SO_2). Volcanoes also emit various amounts of carbon monoxide, fluorine, hydrogen, nitrogen, argon, chlorine, and methane. You sure wouldn't add many of these to your health food cocktail.

Upon reaching the surface the gases cool enough to precipitate various elements and minerals along the edges and around the tops of the fumaroles. Precipitates of iron, copper, zinc, mercury, arsenic, and bismuth, which occur in the gases in trace amounts, give many fumaroles brilliant, splashy colors, making them look like a painters palette; there are creams, browns, oranges, yellows, greens, reds, blues, and grays. Native sulfur commonly precipitates out as soft, beautiful bundles of yellow crystals, called "flowers of sulfur," and precipitates of antimony may form long, steel gray, pencil-like crystals called "blades" of stibnite.

The temperature of fumaroles varies from just over 150° F (65.5° C) to more than 1,800° F (982° C). Higher-temperature fumaroles are rich in hydrogen chloride and hydrogen fluoride, whereas lower-temperature fumaroles (<212° F [100° C]) may contain abundant carbon dioxide. This gas, being heavier than air, can collect in hollows or depressions below fumaroles, or even float down narrow valleys to collect in basins at the foot of the gas god. Animals, bugs, and people who wander into such places can quickly die. This is apparently what happened to a group of berry-seeking grizzly bears in Yellowstone National Park. In a place now called "death gulch," the bears were asphyxiated by carbon dioxide that had traveled down the gulch from a nearby steam vent.[3]

In 1947 carbon dioxide accumulated in hollows at the base of the "entrance to hell"—Iceland's Hekla volcano. Sheep pastured in the hollows "drowned" in the gas, while the shepherds looking after the flocks were not affected. Their heads were above the level of the gas.[4]

Then there was the blue haze of Iceland. Not a singing group or a fraternity initiation, the blue haze was an enormous amount of a deadly, ground-hugging gas released out of a 15-mile-long (24 km) volcanic fissure during and after the great Laki lava flow of 1783. It is estimated that 50 million tons of sulfur dioxide, 5 million tons of fluorine, and 18 million tons of carbon dioxide were released into the air. Carbon dioxide and fluorine are heavier than air so they stayed close to the ground, going where the wind would take them. The combination of these gases created a toxic spill of amazing proportions, a blue haze that killed grass and caused fluorine poisoning (fluoridosis) in animals and some humans. Over 11,000 cows, 28,000 horses, and 190,000 sheep died from fluorine poisoning or starvation, and 10,000 people from the resulting famine.[5] The following winter was harsh all across Europe, prompting Benjamin Franklin, who was then the ambassador to France, to suggest that the gases from the fissure had blocked out enough sunlight to cause the cold weather.

While we are on the subject of pollution, the amount of sulfur added to the atmosphere from big, explosive volcanic eruptions can be immense—thousands and thousands of tons. Enough sulfur can be blown into the stratosphere to form a layer that quickly encircles the earth. This layer, at 10–20 miles (16–32 km) above the earth's surface, can remain airborne for several years since there are no clouds at this level to make the rain necessary to wash out the offending gas. It can effectively block out enough sunlight to cause temperature decreases on the surface and to affect local and possibly global weather patterns. Pinatubo put enough sulfur into the air to block out 2–3% of incoming sunlight, leading to lower temperatures, two cold, wet summers, and possibly the very cold winter of 1994.

However destructive, these kinds of volcanic eruptions do provide us with glorious, breathtaking, beautiful red, green, blue, and scarlet sunsets. For these we can thank the likes of Krakatoa, Pinatubo, Augustino, Leamington, and many more. The ash and gas in the atmosphere disperse the sun's rays, giving us those lovely colors. So in essence there we have it—beauty and disaster, people and volcanoes, nature, Earth, and us.

ENERGY TO USE

The earth, with help from grandma's oven, is a pretty hot place. If this heat is concentrated enough, then it can be harvested and used as geothermal (heat of the earth) energy. Unfortunately, there are only a few places on spaceship earth where this is practical (and economical). By and large we have visited most of these places already, for they are closely associated with the lands of hot springs, geysers, and volcanoes.

Exactly how this "heat of the earth" is used depends on the temperature of the waters below the hot springs and geysers. Where this water is below the boiling temperature (such as the 140 °F [60° C] water below Reykjavik), it can be pumped directly into, and circulated through, homes and business to heat them. If the water is above the boiling point, or very close to it, the waters can be converted into steam and used to run turbines to produce electrical power. At Wairakei, New Zealand, and the Geysers area in California, electrical power has been produced this way for more than 33 years, and at Larderello, Italy, for more than 40 years. The Geysers area has a generating capacity of more than 2,000 megawatts—enough to light half of San Francisco and have enough left over for all the Christmas trees in Minnesota.

Unfortunately, unlike the little engine that had enough steam and could, the Geysers can't. It appears there was never enough water in the rocks to warrant the extensive electrical generating facilities developed there. In most geothermal areas of the world, steam to drive electrical turbines is produced from superheated waters that are brought to the surface by wells drilled into the "hot water reservoir." These wells allow the volume of water in the reservoir to be determined and the maximum generating capacity calculated. The Geysers, however, is all bark and no bite; wells have encountered only steam, no hot water. Steam provides no measure of the volume of water actually present in the reservoir. The Geysers reservoir has been likened to "a boiling teakettle that steams (and whistles) vigorously, until it suddenly runs out of water. The amount of water present is not known, until it is actually gone."[6] Electrical output at the Geysers is projected to fall rapidly over the next few years.

On a worldwide basis geothermal energy, used for generating electricity and heating, saves the world about 55 million barrels of oil each year.

Geothermal energy is cheap and efficient, emits much less carbon and sulfur gases than fossil fuel power plants, and is on tap 24 hours a day regardless of weather, spills, embargoes, or shortages. However, this shining knight in environmental armor is not all pure and sweet. Some of the problems with geothermal power are the following:

1. It has to be used right where it is found. Unlike coal and oil, it cannot be shipped or piped to metropolitan areas, and most geothermal fields are far removed from urban centers.
2. The water (as we have seen from its surface equivalents) can contain many unfriendly elements. If these get into the groundwater system, or into rivers and creeks, they can cause a lot of pollution. This is exactly what happened to the Wairakei River in New Zealand.

3. The minerals dissolved in the hot water can't wait to precipitate out. If they do this within the steam pipes or conduits they will eventually clog them tight. The pipes then have to be taken apart and cleaned out. In New Zealand, however, this is not all bad, because the minerals that clog the pipes are rich in gold, silver, copper, and zinc.
4. Pumping too much water out of the underground system too fast can deplete the water resource. This not only causes the field to cool down, leading to loss of energy, but also allows the ground to crack, subside, and/or sink, signaling disaster for wells, pipes, buildings, and golf courses.

What of the future of geothermal energy? We may have to ask an underground miner to find out. Miners know the deeper down a mine goes the hotter it gets. On average the temperature increases 1° F for every 300 feet (91 m) of depth. The walls of the deepest South African gold mines are at 150–160° F (65–71° C), which calls for some pretty serious air conditioning.

If deep holes (12,000 feet [3,658 m]) can be drilled into the earth's crust (which will make the planet seem much more like Swiss cheese than the moon does), then cold water can be pumped down them to be heated by the natural warmth of the rocks at this depth (220° F [104° C]). This water may then be pumped back to the surface to be used to heat buildings, or turned into electrical power. This could be done anywhere in the world.

Of course there is one pesky problem to overcome. The permeability of rocks decreases with depth. At depths of 2–3 miles (3–5 km) the rocks will be tighter than Scrooge's wallet. In order to get water into and through the rocks there has to be some way of artificially cracking or breaking them. Just how this can safely be done is not known at present (and we certainly won't ask the Army Corps of engineers—their answer to any problem is either build a dam or blow it up).

For now the widespread use of geothermal power remains a cloud on the far horizon. Someday, somewhere, one of us dream catchers will find a way. Until then the earth plays on and the waters simply laugh and dance.

SECRETS OF THE OCEAN NIGHT

"The earth was without form and void, and darkness was upon the face of the deep" (Genesis 1:2). Down in the deep, at the bottom of the sea, dark and cold, but not without form. Down a mile or more below the angry surface amongst the fissures of the midocean ridges. Down where the sun never shines yet life thrives, where no prospector's hit paydirt but great mineral chimneys pour forth clouds rich in copper, zinc, lead, silver, and gold. A strange, strange land unseen by humans until 4 years after Neil Armstrong took his "giant step for mankind."

Unseen, but imagined by the likes of Jules Verne and a British geologist by the name of John Elders. In 1965 Elders proposed that the midocean ridges might be places of hot springs and submarine geysers like Yellowstone and Wairakei. He reasoned that if water could seep through the continental crust, be heated at depth, then sent back to the open air to form a Yellowstone, then the same thing should be happening at the bottom of the sea during the long ocean night.

And we can imagine what it must have been like on that day in 1977, when deep-sea vents were discovered on the Galapagos Rift and Elders was proved to be right. Imagine being inside a cold and clammy submersible, crouching in front of the small viewing window with no room to stand or stretch. You're down some 8,000 feet (2,438 m) in an eerie, windless world dominated by wrinkled, twisted plains of dark lava cracked and split open like overcooked sausages. A world lit only by strange, wispy fish that flash and flicker like summer lightning moving through dark clouds. Suddenly the ship drops into a long, crooked fissure. Slipping around a huge pile of broken rock, the fissure narrows and a steep lava wall looms directly in front. With the skill of a magician the pilot maneuvers the ship up the steep cliff, to the top and over, to a place where time holds its breath and life will never be the same again.

There, caught in the ship's lights, as far as you can see, is a world worthy of Jules Verne's imagination. Milky white water rises from fissures, like smoke from a camp fire, and slowly spreads

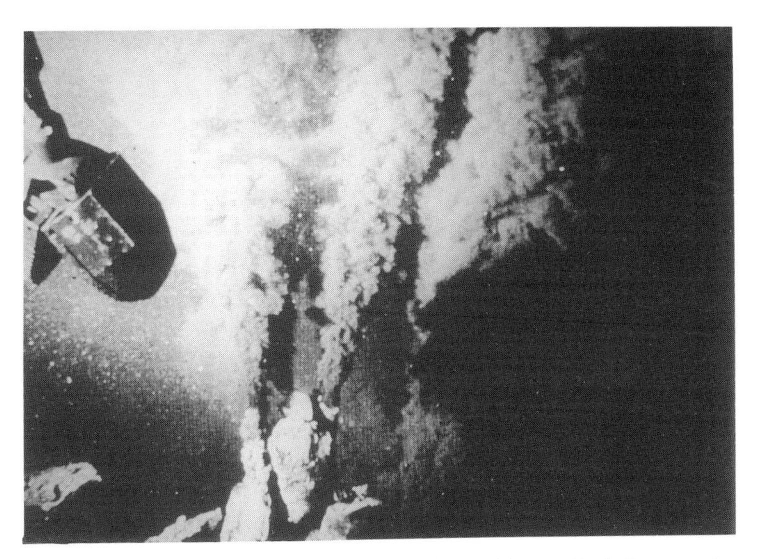

Hot, black, metal-rich water shooting from the tops of stone chimneys. In the foreground can be seen the sampling arm of the research submersible Alvin. (Photo courtesy of J. M. Franklin)

across the cold lava floor to hover and shimmer like some faraway galaxy. Within this island universe stands a freakish forest of stone chimneys, jagged and knobby, brightly colored, shooting geysers of hot, black water out their tops.

Even more wondrous than the ghostly woods are the incredible life-forms. Colonies of foot-long clams grow thick as weeds where the water bubbles forth, and crabs scuttle along the bottom to feed on the swarms of blind, shrimplike creatures that hug the chimneys, or drift silently through beds of blue and white sea anemones. And above all this are the strangest of the strange—grand worms, looking like blood-red tulips, two to fourteen feet tall, blind, groping, and hungry.

Hot springs and life, together, all around the world; but who would have thought it down here, in a place "only a little less hostile than the black reaches of interplanetary space."[7] Down where the sperm whale hunts the giant squid and Captain Nemo tends his gardens.

The midocean ridges: places where jigsaw piece pulls away from jigsaw piece, creating long, deep fissures where hot lava wells up from grandma's oven to fill the cracks and spill onto the seafloor, only to be ripped apart so the entire process can begin anew. These cracks in the skin of the oceanic crust form channelways for gallon after gallon of seawater. Down along these natural aqueducts, the water trickles and seeps and slowly changes as it heats up and interacts with the crispy basalt. From cold, salty Dr. Jekyll into hot, acid, sulfur- and metal-rich Mr. Hyde, who basks in the warmth of grandma's oven, expanding to rise back to the seafloor, to bubble forth as warm-water springs or to gush out the tops of the rocky chimneys like oil from a well.

It's Hyde that gives life and form to the void. Life to zillions of tiny bacteria that bask in the radiant warmth and devour the pungent sulfur that Hyde carries, using it exactly the same way plants use sunlight—to create energy that is used to make sugars and starches. It has been said that these bacteria are the closest living link to the first creatures on earth. The other life-forms in the void either gobble up the bacteria, or spend the dark hours devouring each other.

Such hot spring systems, and their nonpaying guests, have been around as long as oceans have flowed and continents moved. The "discovery of the hydrothermal vents may mean that at last science has found a natural habitat like that at the beginning of the earth."[8]

Do these submarine vents represent the place where life got its start? Is this the primordial soup with all the goodies necessary for the creation and maintenance of life? Here we have heat, turbulence, chemical energy, sugar and spice, and everything nice—all the right stuff, from hydrogen, water, methane, and ammonia to probably most important of all, safety. Freedom from the harsh, poisonous surface where the comets and meteorites play; safety to carry life forward to bigger and better things.

Maybe—or maybe not—
The spinning wheel has yet to stop;
Life began and ventured forth,
Some went south and some went north,
And in the end we all evolved,
But from where we came is not yet solved.

Even if the origin of life is still hotly debated, the discovery of submarine hot springs provided the solution to another great earth mystery, one as old as the Dead Sea scrolls.

And just what do the Dead Sea scrolls have to do with hot springs you may ask? Well, actually I don't know, except the scrolls were found near the Dead Sea, and what do you think makes the Dead Sea dead? Water. Believe it or not it's that giver of life that kills an ocean. Dead seas, lakes, and marshes all have one thing in common: They are trapped between a rock and a high place. Water may run or trickle in, it may rain down, but never, ever, can it get out. Its only escape is to vanish into thin air, to evaporate in the hot sun. When water does this it literally leaves behind the sweat of its labors—the white stuff we call salt. If the amount of water evaporating exceeds the amount of fresh water running in or raining down, then the salt (sodium chloride) accumulates; it makes up more and more of the water that remains, and over the eons, a great salt lake or a vast dead sea is formed.

Now, the rolling oceans have many inlets; water pours in from rivers and creeks and rains down from on high. Not one, however, has its own private outlet. If we total up the surface area of the oceans, if we calculate the average rate of evaporation, and if we figure out the rate at which fresh water is introduced to the seas, it is easily concluded that there should be a lot more salt in the water than there is. In fact, we should all be able to walk on water. So why do we sink and have to swim? Why are the oceans not as dead as a doornail?

"Har, har," laughed Long John Silver, "Polly me pet, it be because of hot springs and moving ocean floors." The oceans are not part of the grateful dead because seawater is continually being sucked down into the oceanic crust to mingle and dance with the basaltic rocks, to change from salty Jekyll to fresher Hyde, to be continually cleaned and purified. In fact, it is estimated that in 10 million years (a mere tick of the geologic clock) as much water as the oceans now hold passes through the oceans' hot springs.

So a mystery is solved, but for every case closed there is another to take its place. For instance, what of those great stone chimneys, some more than 35 feet tall and 2 feet in diameter, and many of them blowing out hot water? Chimneys that often stand on mound-shaped structures that rise several feet above the ocean floor, mounds that have been found to be rich in copper, zinc, lead, silver, and gold. Do the metals occur in economical amounts, and can they be mined? Are they a renewable, economical resource, much like reseeded forests, with the mounds continually forming and growing as seafloors spread and hot water rises? And finally, what of the rest of the midocean ridges—the 95% that have yet to be explored? What new discoveries await down in the void, down in the rifts and fissures, just beyond the searching lights, and around the next bend?

Prelude to Chapter 6

Silence, eons of quiet, as concentrations of metals slowly form within the earth. In contrast is the noise that accompanies the coming of the miner: from the sound of a metal pick striking hard rock, to the din and vibration of heavy machinery drilling, digging, and loading.

The music starts with the tinkling of a xylophone, constant and steady. Introduced into this is the strong and regular rhythmic pattern of the string instruments as the miners come and dig and take away. Last is a slow decline, the metal is mined out, and the tinkling xylophone falls silent.

CHAPTER 6

Harvesting the Earth

From the dancing waters
We harvest a wonderful crop,
Element after element
That lets our civilization grow, fly and hop,
From copper and gold to lead and zinc,
Making everything
From airplanes right down to the kitchen sink.

Take computers, for instance, those electronic machines we love to hate. It takes more than 30 elements just to make the darn things go "beep." Out of this gang of 30, geothermal waters carry, precipitate, and concentrate copper, zinc, lead, mercury, gold, silver, tin, antimony, gallium, tellurium, molybdenum, and barium. Thirteen out of 30—that's good enough to win the American League batting title.

However, some 6,000 years ago, a mere twitch in geologic time, the elements that started humans on the road to civilization—copper, gold, lead, and tin—weren't even in the top 40 of the "useful things" list. No, back then the real hot commodity was something called flint.

What Stone Age humans called flint came in two forms: a mineral that was a very fine-grained variety of quartz (SiO_2) and a dark, glassy volcanic rock called obsidian. The mineral and the rock quickly moved into the number 2 and 3 spots on the list of useful things, right behind that all-time cave favorite—fire. Flint and obsidian were all the rage because they could be chipped or "flaked" into sharp-edged, shell-shaped pieces perfect for spear tips, arrowheads, and scrapers of all sorts, including the world's first razors.

The world's second industrial mineral after salt, flint, for all its hardness, had several inherent weaknesses. It turned out that flint was brittle and broke easily, dulled rather quickly, and was a pain to work with. So flint, though it put humans on the road to success, was soon replaced by bright, shiny metals.

THE RED METAL

Of all the metals gold was probably the first to attract human attention. Bright and yellow, glittering amongst the whiteness of the stream gravel, it was hard to miss and even harder not to pick up. Once found and messed around with, it probably didn't take humans long to figure out that this soft, pliable substance could be shaped and hammered into plates and thin strings. Great for decorating the cave, or yourself, but gold was pretty much useless as far as making the really important stuff, like tools and weapons.

Far more valuable than yellow gold were lumps of a soft, red metal called copper. Nuggets of native copper could be shaped and hardened by hammering to make the sharpest of tools, the slickest of knives, the choicest of spears.

Alas, poor flint, we knew you well, but now you've gone the way of glass milk bottles and ice boxes. Copper was the new lord of the cave, but like flint, it too had a potential fatal flaw. It turned out that if you hammered the red metal too long, it passed the point of hardening and turned brittle. Imagine the proud hunter, his breath forming a thin cloud in the chill of the early morning as he

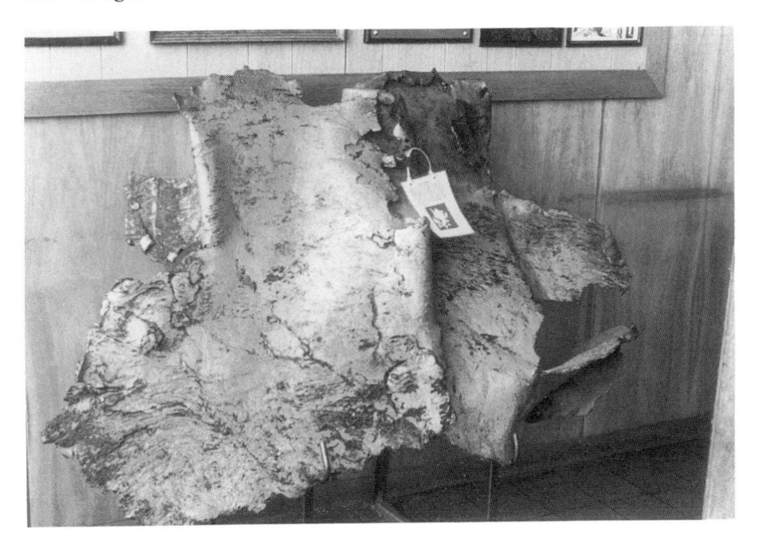

Large "nugget" of native copper found in the White Pine Copper Mine enclosed by layers of shale (slate). The mine is located in the upper pennisula of Michigan. (Photo by P. Morton)

thrusts his shiny, copper-tipped spear into the exposed belly of the big animal. Slicing through the thin fur it enters the tough skin and the unthinkable happens—the tip shatters into many splendid pieces. And you think you got problems?

The solution to the brittle copper problem was probably discovered by a group of teenagers standing around the fire trying to act cool. Listening to the rhythms of a hollow bone flute and hide drum, one of them picks up a lump of the red metal and tosses it into the fire. Now that seemed pretty neat so they all tried it. But after a couple of handfuls they went back to talking of great hunts and soft, warm robes. Suddenly one yells and points into the flames. They all stop and stare. The copper is bright red, it's glowing, and then, as if by magic, it turns into flowing liquid.

Can you imagine the yelling and shouting, the running to get the others? And while they ran screaming into the night, the copper ran in rivulets across the rough ground to fill small holes and depressions. When the fire was doused to a smolder the entire clan stood astonished. There before them was the red copper, but it had formed new, smaller lumps the exact size and shape of the holes it had filled. This was indeed magic.

Not only could copper be heated to make it softer and more workable, but when heated enough it melted and would then take on the shape of anything it ran into. Thus was born the idea of making molds or casts, a mold of an axe head to be filled by liquid copper, or an arrowhead or even cave baby's first deerhide shoes. Another benefit of this discovery was that heated, softened copper, when cooled, could be worked without becoming brittle.

After this it didn't take humans long to figure out that copper was just like the H. J. Heinz company—it came in 57 varieties. Copper was not only red but also green, blue, black, gray, and purple. When copper dressed up in these other colors it was to step out with other elements such as sleek carbon or sophisticated oxygen. Together they made new minerals like blue azurite, green malachite, purple bornite, and black chalcocite. From teenagers throwing copper lumps into the fire came the idea that heating other minerals might restore copper to its native state. Thus the art of smelting came into being, and overnight the amount of copper available for tools and weapons increased substantially.

The discovery of smelting minerals to get metals out occurred around 4,500 B.C. in Turkey and Iran and also, at about the same time, in the great North American wilderness at a place called the Keweenawan of Michigan. Here native Americans mined not only native copper but also green (malachite) and blue (azurite) copper minerals. They smelted the copper metal out of these colored minerals by heating them in a hot, charcoal fire.

As copper went, so went the Pharaohs of Egypt or, more importantly, those great monuments to themselves—the pyramids. The early rulers of Egypt had to settle for second-rate monuments. They had no tools with which to cut and shape anything but mud and clay, so all they could afford were low, ugly buildings constructed of mud bricks. The coming of copper changed all this and led to the building of the stone pyramids. The largest pyramid of all is one of the three great pyramids of Giza. It took 2.3 million limestone blocks, weighing an average of 2.5 tons, to construct it; the heaviest stones weighed in at 45 tons. All of these were cut and shaped with copper tools (how they moved the blocks from the quarries and hoisted them into place is another story entirely).[1]

Copper was indeed queen of the Nile until about 3000 B.C. when some no goodnick in Iran added a pinch of a white metal called tin to the liquid copper. The tin had been burned (smelted) out of a ruddy-colored mineral called tinstone (cassiterite).

Faster than priority mail, copper had a partner and the Bronze Age was upon us. That little pinch of tin made the copper much harder without making it brittle. Humans could now hammer, hack, slice, hew, cut, and mangle to their hearts' content—the world's first alloy (a mixture of two or more metals) had been made. So bronze became "tougher than copper," and for the next 2,000 years would be hailed as the "cornerstone of world industry and art" (until it gave way to iron).

Gold for decorations, copper and tin for tools and weapons, and lead for—lead? Just a minute here. How did that dull, heavy, worthless stuff get in with this distinguished company? Alas, poor lead, unwanted and unused until someone heated it in a fire (sure seems like the march of metals is ever directed by the flames of a hot fire), where it sizzled and was no more. The lead disappeared in a thin white vapor, almost like steam from a boiling kettle. But sitting in lead's place was a small, shiny, white drop of pure silver. Lead to silver, surely this was an alchemist's dream.

But how? Certainly we don't believe in alchemy or magic—do we? Well in this case the answer is no, of course not. It turns out that the elements silver and lead have very similar chemical properties and so they have a natural affinity for each other. During the formation of lead (generally as a lead sulfide, a mineral called galena), any bits of silver hanging around will be incorporated into the lead's crystalline structure. When heated the lead becomes quite volatile, like a mother with four children finding out their flight has just been canceled, and vaporizes, leaving behind that which it originally ingested—silver. This process of getting silver from lead is called *cupellation* (a cupel being the porous, cuplike dish that is used to hold the lead); this method was to represent a major source of silver for ages to come.

So lead begat silver and silver begat shekels, which were humanity's first, widespread standard of exchange. A shekel was a quarter-size piece of silver (about ½ oz.) used by Babylonian, Phoenician, and Hebrew merchants as a unit of weight (1 shekel = 18 minas, and 60 minas = 1 talent). Also, at about the same time, it was established that 1 unit of silver was worth 180 units of copper, 40 units of lead, and $\frac{1}{6}$ of a unit of gold. Silver and gold were on the rise, while copper and lead were on the outs.

Much of the copper that built the pyramids, the silver of shekels, the gold of cave debutantes, and the tin of weapon makers was a gift from geothermal waters—Mr. Hyde slowly bringing his treasure to the surface for humans to find and use. Turns out Hyde isn't such a bad chap after all. Humans have been "mining" Hyde's treasures for around 10,000 years, whereas old Mr. Hyde has been busy making them for as long as oceans have had tides and continents shores.

BY THE LIGHT OF THE SILVERY MOON

We humans have always been a curious race. We are forever wanting to know the whys and wherefores of all kinds of happenings and natural events. The formation and concentration of minerals, and metals in particular, has been no exception. Ideas, concepts, and guesses on how the metals got to where they are and from whence they came have been as numerous as gators in a Louisiana swamp for at least 3,000 years.

For instance, long ago, when the earth was the center of the universe and the sun a mere planet, it was believed by many "educated" folk that a great fire raged at the center of the earth. This fire was a real holocaust that emitted clouds of metal-forming vapor, which rose to the surface where it cooled to form minable metals. People of a more "scientific bent" thought this was nonsense. They claimed the metal-rich vapors came from the subterranean combustion of coal and sulfur, which was ignited by the heat of lava rising to the surface to build volcanoes.

The alchemists had a much better idea, one that actually held sway for several hundred years and still lingers on today in some of the expressions we use. They believed that the sun and planets and moon were responsible for the formation and concentration of the "useful" metals. Each planet developed its own special metal and, to a certain extent, gave the metal its own rare and peculiar characteristics.

The sun, being bright yellow, directly influenced the formation of the yellow metal gold. For this reason almost all gold would be found in southern climates, for it was here that the sun's "golden" rays fell most intensely. Many people, including scholars and the leaders of Spain and Portugal, said it was foolish to search for gold in northern climates—plain silly to get more than a stone's throw away from the equator.

The idea of the sun's role in gold formation ran so deep that it actually played a part in the exploration and settlement of North America. On a foggy, rainy day, some 400 years ago, the British fleet, under the command of Captain George Vancouver, met the Spanish fleet off the west coast of Canada. This meeting caused an immediate dispute over which country had the right to explore this vast, mountainous land. Even considering there was no overnight mail, the dispute ended rather quickly. The Spanish had no real interest in what was to become Canada, and thought the British a rather stupid lot for even caring. "The indifference of Spain to this vast, unexplored territory was explained by the preoccupation with the quest for gold and the commonly held theory that the nearer the equator you are the better the chance of finding it."[2]

What of other elements and their planetary origins? The alchemists also believed that the growth of silver was influenced "by the light of the silvery moon." The planet Mercury, named after the fast-moving messenger of the gods, gave us quicksilver, or the element mercury. The god of war, Mars, supplied iron, and the cold, distant world of Jupiter was responsible for the heavy, dull element lead.[3]

Since the days of the alchemists we have indeed come a long way. Modern theories on how the metals came to be and got to where we can get them are now much more rational and scientific. Take the theories on the formation of "useful elements" by geothermal waters. Today we know this happens because Dr. Jekyll likes to become Mr. Hyde (how much more scientific can you get?). Here is the recipe. Start with cold, saline, anemic seawater. Add a lot of heat, stir, and slowly percolate through the upper crust (this slow brewing and natural heat aging in basaltic lava leads to a truly natural blend). After a few tens of years, tap at the surface and you will find a hot, acid, metal-rich fluid, that is now perfect for metal making. The sun's gold, the moon's silver, Mercury's quicksilver, Jupiter's lead, Venus' copper, and Mars' iron have been slowly leached out of the volcanic rocks by the passing waters.

THERE'S MAGIC IN THOSE WATERS

Mr. Hyde's waters are able to travel through the earth's crust, collect rare and diffuse elements, bring them to the surface, and with a wave of a geyser, precipitate and concentrate them so we can come along and scoop them up.

How rare and how widely scattered are the elements on whose shoulders the civilized world rides? Imagine the New Orleans Superdome filled to the brim with colorful balloons. There are blue ones for the element silica, white ones for aluminum, green ones for iron and magnesium, gray ones for calcium, pink ones for potassium, black ones for sodium, and orange ones for phosphorous and manganese. There are lots and lots of blue and white balloons (it looks like a political convention) and a lot less of the green, pink, black, and orange balloons, but enough to be easily seen. Together these balloons fill up 98% of the Superdome. The balloons that represent the elements we are interested in can't be seen at all—not a hint nor a glimmer. If we were aloft in the

Goodyear blimp and able to use a powerful and high-resolution camera (which is also an X-ray camera, for it lets us see right through the balloons), we just might be able to pick out a small, red balloon over by the goalpost. It is sandwiched between three blue ones and a green one and covered by a zillion more balloons. This red balloon represents the element copper, which makes up a whopping 0.006% of the balloons in the stadium (and the same amount in the earth's crust). Over by the home team's bench we just might spot a quarter-size silver balloon representing the element tin, which makes up 0.0002% (2 parts per million) of the balloons in the stadium. Finally, right on the 50 yard line, sits a pea-size yellow balloon—this is gold—which makes up 0.0000002% of all the balloons present. Tiny, scattered from one end of the Superdome to the other, yet the slowly moving waters, like the best of bloodhounds, have the uncanny ability to sniff them out, dissolve them, carry them to the top of the building, and present them to the pilot of the Goodyear blimp. This is indeed magic, and it is done without smoke, mirrors, or cameras.

In general, the amount an element needs to be concentrated in order to make harvesting worthwhile is inversely proportional to the element's average abundance in the earth's crust. Abundant metals, such as aluminum and iron (at 8% and 6%, respectively), need to be concentrated only 4–5 times; good old copper, at 0.006%, needs to be concentrated 100 times; tin, at 0.0002%, needs to be vacuum packed and concentrated at least 2,000 times; whereas an element like mercury, at 0.000002%, needs to be scavenged and concentrated 100,000 times.[4] This is just what Mr. Hyde is capable of doing (and a good thing for us too).

Exceptions to this general rule of thumb (and in geology there are exceptions to everything—including the exceptions) are those very few elements that are so highly prized or priced that just a small quantity makes them minable. Such elements include gold and platinum (where you need about 10 parts per million to mine them) and gemstones, where one carat amongst several tons of rock makes it a minable commodity.

The hot, metal-rich waters of Mr. Hyde live in the same neighborhood as bathtubs, teapots, and turtles. Currently active and once active geothermal systems (fossil ones) are most commonly associated with volcanoes and earthquakes. Places where metal-bearing fluids are presently bubbling away include Japan, New Zealand, Indonesia, the Philippines, Alaska, Yellowstone, and long, snaking lines called midocean ridges. Over the past 3.5 billion years volcanoes and hot springs have come and gone, but in many instances Mr. Hyde's treasure has remained buried in the rocks, waiting for someone to come along and unearth it.

THE GLASS HOUSE THAT METALS BUILT

Resources are like air—of no great importance until you are not getting any.

On a geologic time scale today's volcanoes and geothermal systems represent tomorrow's minable mineral crops. On a human scale we consume minerals and elements much faster than the earth can grow them. In the first 50 years of the twentieth century, total production of minerals and mineral fuels was greater than the total production during all of previous history; in the last part of the century we have already mined more than in the previous 50 years. Very simply, we are devouring the earth's treasures like sharks on a feeding frenzy.

For example, within the house of metals called the USA lives 6% of the planet's human population, yet this place uses 33% of all the minerals and mineral fuels dug, pumped, processed, and packaged each year. In terms of consumption, each one of us Americans, regardless of age, race, or number of TV sets, accounts for 1,300 pounds of iron and steel (and many steels use elements like nickel, molybdenum, vanadium, chromium, tungsten, etc.) (590 kg), 65 pounds of aluminum (29.4 kg), 25 pounds of copper (11.3 kg), 15 pounds each of lead, zinc, and manganese (6.8 kg), 8,000 pounds of oil (3,628 kg), 4,700 pounds of natural gas (2,131 kg), 0.1 pounds of uranium (0.04 kg), and 5,500 pounds of coal (2,495 kg).[5] As our population grows we will be faced with two choices. Either we mine more and/or import more to maintain the status quo, which represents our current standard of living, or we lower our standard of living (not a popular choice). Mine more or live less—that's what the earth might call civilized suicide.

Add to this the fact that our minable reserves are declining and you got trouble in River City, my friend. Mineral reserves are defined as the amount of a given mineral that has been found and can be economically used with existing technology. The term is also restricted to minerals that we have not already devoured. According to government figures (Mineral Commodity Summaries, 1994), current world reserves for such elements as tin, zinc, silver, lead, copper, gold, and sulfur is less than 35 years.[6] What Mr. Hyde toiled 3.5 billion years to create, we've used all but 35 years of.

Well, you know governments and their reliable statistics; world reserves are not as grim as all this. The above kinds of estimates are as conservative as Barry Goldwater, Pat Buchanan, and Rush Limbaugh combined. As economic and technological factors enter into reserve calculations (such as price increases, increased mining technology, and new discoveries) we can easily add many years to the reserve base. But still there is an end, and unlike the pot of gold at the end of the rainbow, this one is definitely in sight.

In our self-defense many of us have started to lend a helping hand by recycling. Recycling will add years to our reserves because it reduces the need for additional production from our reserve base. We aren't doing too badly here; as of 1994, we are recycling 64% of our lead, 40% of our zinc, 24% of our copper, 27% of our tin, 20% of our silver, and 26% of our nickel.[7] Now that's a long way from perfect, but it's one big frog jump better than it was 5 years ago.

Over the past 25 years the techniques used to explore for, and find, new concentrations of valuable minerals have improved dramatically. Geologists that explore for mineral deposits use many varied techniques, such as satellite imagery, infrared photography, electrical conductivity of rock bodies, density of rocks, magnetic intensity of rocks, chemistry of rocks, and, finally, the most important tool—good, sound geologic field mapping or "reading of the rocks." With all this good science and great technology we should be finding mineral deposits about as often as universities raise tuitions. Unfortunately, the rate of discovery has actually decreased over the past 35 years, almost as dramatically as the Canadian conservative government slid to oblivion in the 1993 election.

This dramatic demise holds true for every commodity but gold, and 35 years ago, with gold at $35 per ounce, no one was looking for it. With the freeing of the gold price, exploration skyrocketed and many new deposits, especially in the western United States and the Pacific Rim, were quickly found. It has been just the opposite for the rest of the world's metals.

As an example of the relative decrease of mineral discoveries we can look at one of Mr. Hyde's premier deposit types. These are concentrations of zinc, copper, lead, gold, silver, and barium that look very similar to the mounds and chimneys now forming in Davy Jones' locker under the watchful eye of Captain Nemo. Deposits on land are called "massive sulfides," and from 1960 to 1980, the number discovered would be enough to make several symphony orchestras; from 1980 to 1995 we're down to two string quartets and a mariachi band. All this high powered technology to find diddly-squat. Most geologists in exploration are dedicated, hardworking, and smart. They are out there doing their best to discover new mineral resources. So if the problem is not the searchers, then what is it?

"Ah yes," spoke W. C. Fields, "Perchance the problem lays hidden under all those layers of well-wrapped charm?" Or, in geologic terms, we just don't know how to explore for mineral concentrations that are hidden under a few hundred to a few thousand feet of hard, dense rock. Most of the mineral deposits discovered from 1960 to 1990 were either exposed right at the surface or buried by a few feet to a few tens of feet of rock or dirt (called overburden). It is now indeed rare to find a mineral deposit "outcrop" (exposed at the surface), or even one located within a couple hundred feet of the surface. Sure, there are lots of underground mines, and many of these extend for thousands of feet below the surface. But the deposit being mined was originally discovered right out in the crisp, clear air and has since been followed down into the bowels of the earth.

Today's explorers are searching deeper and deeper for mineral deposits and so far the earth is winning. I think most geologists would agree that the deposits are down there. It would indeed be an extremely cruel joke if the planet placed all of its mineral treasures, developed over more than 3 billion years of earth's history, within the first 100 feet of the surface. So what it comes down to is this: just as the three blind mice were having trouble with the farmers wife, so blind geologists are having trouble seeing the yellow, red, or silver balloons down at the bottom of the Superdome.

Another factor that has led to our poor record on cutting the mineral deficit is that minerals are where you find them: they are not equally distributed around this blue planet of ours. The earth's mineral crops have no national anthem, no allegiance to artificial subdivisions of the land. Geologists just can't go anywhere and find copper, silver, lead, or whatever. Unlike Quaker Oats, minerals are not shot from guns and scattered evenly around the earth. It is just the opposite. Mineral crops occur within specific rocks at specific places in the earth's crust, and they do so for strictly geologic reasons.

This unequal distribution of mineral resources has led to some pretty hard feelings over the years. Here's one country absolutely rolling in valuable "goolydozzle," while its neighbor has been blessed with none. Such situations cause a proliferation of border disputes, aggression, colonialism, wars, and economists. It is because of this unequal distribution that the United States started what is now called the "stockpile of strategic resources." This began during World War I when Germany decided to stop selling us several critical minerals (such as potash and salt). Today the term *strategic* refers to minerals necessary for energy programs, industrial maintenance, and military defense. These now include chromium, cobalt, nickel, platinum, titanium, manganese, aluminum, fertilizers, oil, and gas. Many of those that occupy the top 10 chart are virtually nonexistent in the rocks that underlie the land of the free. We import most of our chromium, manganese, cobalt, platinum, and titanium from South Africa, nickel and potash from Canada, aluminum from New Caledonia, and we all know about the "Texas tea" that we import from the Middle East. Having a strategic stockpile is obviously a good idea.

The definition of a strategic resource has varied greatly over the centuries. For early humans strategic minerals were salt and water; for the Romans they were iron (for weapons), gold (to pay their soldiers), and lead (for pipes to carry their water). In today's complex society the list has been dramatically lengthened. Other than the top 10 cited above, strategic resources now include such things as frozen pizza, VCRs, disposable diapers, and an instant cash card.

Unequal distribution of minerals means that geologists are limited as to where they can search for specific mineral deposits. For instance, we have seen how deposits of copper, zinc, lead, gold, silver, and barium have a great affinity for volcanoes and rising hot waters. To find these deposits, today's explorers must look in areas that are either volcanically active, or were once the lands of mighty gods. On the other geologic foot, nickel, platinum, cobalt, and chromium deposits are inseparable from large mafic intrusive bodies like the Duluth Complex in Minnesota, the Sudbury Irruptive in Sudbury, Ontario (the nickel capital of the world), and the Bushveld Intrusive in South Africa. Diamonds occur only with pipelike bodies known as kimberlites, very special rocks that are found only in a few places around the world. In exploring for specific mineral commodities geologists have to search within the "right stuff"—the right rocks, the right environment—the right place. If it turns out that it's politically or financially impossible to look in the right places (either because of taxation policies, unstable governments, environmental policies, etc.), then the mineral is just not going to be found. Given the current rates of consumption and discovery of new deposits, we may one day face the task of finding usable substitutes or of having to reduce our standard of living.

To some a reduction in living standards equates to what they nostalgically call "the good ol' days," our pastoral past, when we hauled water, hewed wood, and dug outhouses. The good ol' days were probably much like Dixie Lee Ray writes about in *Trashing the Planet*:

> The world in which I spent my early years was a very smelly place. The prevailing odors were of horse manure, human sweat, and unwashed bodies. A daily shower was unknown; at most there was a Saturday night bath. There were no electric refrigerators or freezers. Food was kept in a "cooler" or ice box, there were no electric irons or toasters and no vacuum cleaners. We used brooms, and carpet beaters and muscle. There were no electric washing machines or dryers, only tubs that could be heated on the kitchen stove.[8]

We can also throw in no indoor toilets, no telephones, frozen foods, and few medicines. I really think very few people want to return to the good ole days. If that is true, then to maintain and improve our lives we either grow it or we mine it, and if it looks like we're going to run out of it we have to put the time and money into finding acceptable substitutes. It's as simple as that. We humans always like to think we are masters of our own destiny. But if there happens to be a predestination in this world it is surely in the form of geologic events into which all nations, miners, environmentalists, and politicians must fit. If we can't or won't, then we will go the way of the dinosaur, and the earth will give a jolly wink, replace our section of the orchestra, and spin on through the void.

OF KINGS AND QUEENS

Of all the minerals and elements associated with volcanoes and volcanic rocks, two are the talk of the town, gold and diamonds— the prized and the possessed. These royal elements have fired the imaginations of storytellers for centuries, determined the worth of monarchs and nations, measured the extent of human greed and jealousy, and plumbed the depths of the human spirit.

Metal and jewel have instigated wars and started frenzied stampedes to the remotest places on the globe, and yet each, over the years, has achieved vital importance in modern industry.

Just thinking about gold and diamonds can send icy fingers trailing down your back. Remember the tales of lost gold and diamond mines, pirates' treasure and laden Spanish galleons gone to the bottom of the sea. I know for sure, out there somewhere, treasure chests are piled high, hidden in the dark depths of a secret cave guarded by swarms of red-eyed snakes. Out there, somewhere in the high mountains, are rocks that shine as bright and yellow as the golden sun. Jewel and metal, fact and fantasy, brings out much that is good in us, and much that is silly and downright evil.

THAR'S GOLD IN THEM THAR HILLS

> We wanted the gold and we dug for it,
> Worked and lived like the lowest of slaves,
> In the end a few of us found it,
> But most only dug out our graves.

Precious gold is said to be a noble metal. Gold is called precious and noble not because you treasure it like Scrooge did his money, or because it is dignified and grandiose. No, it's called precious because, even for a metal, it is rare, and it is noble because it resists corrosion and is therefore able to stand up to just about anything.

Soft and malleable, gold has such little tendency to corrode that "most of all the gold that has ever been mined is still in use."[9] The gold in your gold crowns may have come from Solomon's mines, and the gold in your ring may have been dug by a forty-niner. Once gold has been mined it can be melted and remelted to be used over and over—rings to watches to coins to earrings and back again.

In all of history we have mined about 112,000 tons of gold. If that were all stacked in one place it would form a cube measuring 58 feet (18 m) on a side, smaller than your average city lot. However, what it would lack in size it would make up for in weight. A 1 cubic foot block of gold weighs 1,185 pounds (538 kg), which is more than two sumo wrestlers with four feet; you can imagine what a 58-foot cube would weigh.

Most of this gold once belonged to Mr. Hyde and has been (or is now) dug out of fossil geothermal systems or deposits called placers. In placer deposits the gold is freed from the surrounding rocks by surface weathering, and then concentrated by running water. Concentration occurs because gold is heavy (it can be naturally separated from other, lighter minerals), resistant to chemical attack, and durable; this means gold will end up in the same places in a moving stream. Other minerals that form placer deposits are platinum, native copper, diamonds, magnetite, and chromite.

Gold in placer deposits occurs concentrated in what prospectors called pay streaks or white runs (white because the stuff of gravel, white quartz, is typically found with the gold). Pay streaks are narrow and rich. They also may be irregular in outline; they may branch, split, or even be absent in parts of the run. Such an uneven distribution can easily lead to bad feelings. Here you are with hardly a speck, and right next to you is some yahoo pulling nuggets out the size of plums.

It was the ease of working gold placers (as well as the value) that caused the great gold rushes of modern times. The placer miner needed but a shovel, a pick, and a gold pan (as well as eyes in the back of his head) to extract gold and in a fortnight, to become richer than your average pro athlete. In Barkerville, British Columbia, in 1868 a heavyset man by the name of Henry Diller dug out 102 pounds (46 kg) of gold in one day, and where the Eldorado meets the Bonanza, in a place called the Klondike, a single pan of paydirt yielded $1,000 (at $20 per ounce). The claim's owner, an ex-mule skinner turned prospector by the name of Dick Lowe, immediately got drunk and from all reports stayed that way.

Most placer gold occurs as small specks called dust, but to the joy of many a miner larger lumps called nuggets may also be present. Nuggets range from pea and bean size all the way up to eggs and baseballs. The largest nugget ever found was in Australia and was called the Welcome Stranger; it weighed 2,280 troy ounces (130 lbs, or 59 kg). The largest nugget to come out of the Klondike weighed 85 troy ounces (4.84 lbs, or 2.18 kg) and the largest one from Brazil weighed 13.7 pounds (6.2 kg). At the other extreme is flour gold, minute, scaly particles of gold which float easily and therefore can travel down the entire length of a stream. The extreme fineness of this gold gives the gold prospector the "colors" in his pan.

It was the gold rush to California that started it all. In 1848 Jamie Marshall discovered gold at Sutter's Mill. By 1849 Sutter's Mill was Sutter's town, with a population of 4,000 forty-niners. The diggings were so crowded that newcomers had to go miles to find an unclaimed piece of ground to work. But the gold was there, and it was not unusual for a claim owner to make $500 a day—not including the can of nuggets. In the year of the forty-niner more than 150,000 pounds of gold (68,100 kg) were scooped out of the ground. But it wasn't only the miners making money. Most commonly those who got the richest were the ones who sold the miners the pans, shovels, lumber, and other supplies. One such man, seeing many of the miners with ragged, torn, and worn-out pants (and many wore their pants until they were shreds because of the outrageous price of laundry), got the idea to make a more durable, longer lasting pant out of canvas. So here it was that Levi Strauss made and sold the first pair of Levi's.

In 1850 there was a small rush to Australia, but the mob scene was waiting in the wings. Forty-six years later the news finally filtered down from Alaska to the lower 48, and the world's wildest, craziest, gold hunt was underway.

In 1896 George Carmack, a man who read *Scientific American*, wrote poetry, and cared more for the color of a salmon than the color of gold, struck it rich on a creek he called Bonanza.

"I took the shovel and dug up some of the loose bedrock. In turning over some of the flat pieces I could see the raw gold laying thick between the flaky slabs like cheese sandwiches."[10]

Bonanza Creek ran into the Klondike River, the center of gold fever. The rush to the Klondike ranged from the heights of stupidity (a group of New Yorkers tried to bicycle to the Yukon), through the

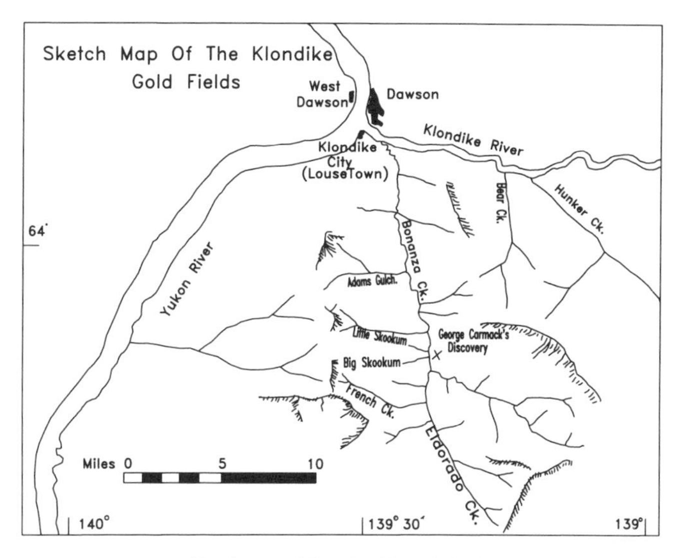

Sketch map of the Klondike goldfields.

slopes of silliness (an Englishman went just for a jolly good time and took his lawn tennis set with him), to the depths of tragedy (people died trying to get there and the miners lived with scurvy and lice and near starvation). But from start to finish it was the people who made it what it was—over 40,000 in the town of Dawson in 1898, and another 40,000 on the trail to Dawson.

To get to Dawson and the goldfields you had two choices. Which you took depended on your "poke" and how much of a hurry you were in (most weren't in a hurry, they were in a frenzy—like football fans trying to get down to the field after the big game). First there was an all-water route from Seattle to a place in the middle of nowhere called St. Michael's, which sat on the edge of the Bering Sea near where the Yukon River emptied into it. Then it was up the mighty Yukon to Dawson, a grand total of 4,200 miles (6,758 kms) for which you could receive up to 1,500 free frequent steamer miles, good toward a round-trip fare anywhere in Canada.

In good weather, with luck, you could make the trip in 40 days; for many it was bad weather and 8 months.

The second way was cheaper, more direct, and much more backbreaking. You would leave the same way, by boat from Seattle or San Francisco up the coast to a mud wallow called Dynea on the Alaskan coast. Then it was up and over Chilkoot Pass, across four lakes and several rapids to the Yukon River, and on into Dawson.

To get to Dawson from Dynea you had to cross into a country called Canada. Canadian authorities were fearful more people were getting into the territory than the land (wildlife) could support. So they enforced the rule that each and every person had to bring in supplies to last one year. That comes out to a whopping amount of stuff to get from Dynea to Dawson. I don't know about you, but my feet ache, my back hurts, and my hands are cold just thinking about it. For there was only one way to get it there—carry it: the flour, the coffee, the cast iron stove, and the port-a-potty (so what if it was just a wooden bucket—every little bit of weight counted).

If you were flush with cash after buying 2,000 pounds (906 kg) of groceries and other odds and sods you could hire Chilkat Indians to pack your goods the 27 miles (43 km) from Dynea up and over Chilkoot pass and down to Lake Linderman. For this shuttle service they charged 38 cents a pound and could carry 200 pounds (90.6 kg) at a crack.

Most greenhorns certainly couldn't afford that, so they packed their own goods, which generally meant making as many as 20–40 trips over the pass, taking up to 3 months! Now, we're not talking here about a stroll up a grass-covered hill with a basket of salmon and cucumber sandwiches and a bottle of Chablis. What we're talking is hiking up a steep, icy, windswept, rugged, mountain pass carrying as much weight as you can plus a few pounds more; you're a human mule with only one task—put one foot in front of the other for 27 miles.

This sure isn't my idea of a trip to a theme park, but the Chilkoot roller coaster had a leg up on Disney. People were lined up for miles to get over that pass. It's been said that if a person stepped aside to rest they might have to wait hours before they could get back on the trail again. Many took one look, abandoned their goods then and there, and fled south. These may have been the timid but they were also the smart. The more hardy struggled and toiled to Lake Linderman where they built wooden longboats to transport their ton of goods across the lakes, through the rapids, and on to the Yukon River.

Lakes Linderman, Bennet, Marsh, and Tagish were right in the worst mosquito and black fly country in the entire territory. The insects were thick as snow in the Sierra Nevada for 24 hours a day, and the trekkers had no Off, Raid, or bug zappers. "Klondikers came to welcome rain (and cold) as an insect repellant and took to inventing stories about Yukon mosquitos that carried off eagles as food for their young."[11] By the time most of these weary, bug-hating adventurers arrived in Dawson, the good ground was claimed, the gold was freely flowing, and all that was left was to labor for wages on another's gold strike.

In 1898 there were "ten thousand men at (Lake) Bennet, ten thousand at Linderman and 20,000 shuttling their goods up the trail from Dynea and Skagway."[12]

What of the sourdough and the greenhorn who already had their claims on such creeks as Bonanza, Eldorado, French, and Little Skookum? In general it was said "any man who spent a summer shoveling gravel and sticky red clay with mosquitos (and black flies) biting 24 hours a day and subsisting on beans, dried fruit and sourdough bread knows all there is to know about hell"[13]—and lotteries. The topography of the area east and south of Dawson—the gold country—was deceptive. The ancient streams that had carried and concentrated the gold were long gone, and it was difficult for even the expert gold hunter to read their old routes. As it turned out this was one gold strike where the experienced eye of the veteran sourdough had no advantage over the greenhorn's. Add to this the 5–25 feet (1.5–7.6 m) of wet, icy permafrost that covered the gold-bearing gravel, and you had yourself one fine guessing game—"you're more likely to find it where it ain't than where it is" was a common lament.

George Carmack initially struck it rich with what he found on the surface, but this gold was only good enough to turn a smelly, lice -infested prospector into a proper gent. To find enormous riches that would make King Midas drool, the claim owner had to dig (or melt his way) through the cold, wet permafrost down to bedrock. His prayer, day after backbreaking day, was that just above the bedrock would be a layer of white gravel, speckled with yellow fire, marking the course of a once golden stream. If it did—he was a rich man.

Louis Rhode was feeling like a total fool as he dug down through the permafrost. All his neighbors were laughing at him for staking such a worthless piece of property, so obviously out of the gold run. He'd wanted to leave the Klondike, go home, but he couldn't sell his claim for enough to get home on; no one in all of Dawson wanted his ground. So with winter coming and nothing better to do, he decided to be stubborn and dig to bedrock. Several weeks later, filthy and bone tired, his shovel clanged off something solid. Digging away the permafrost he brought his lamp closer and there, on top of a pinkish colored rock, was pure white gravel streaked with flakes of gold. When you started you just never knew. Rhodes took out over $500,000 from his "worthless" claim.

Then there was Anton Standler, a 29-year-old ex-cowboy and coal miner from Austria, who found a little trickle of a stream on what was called the "South Fork of Bonanza," a tributary of a tributary that had been ignored by veteran prospectors. From his first pan he took $6 worth of coarse gold grains. He and his four companions paced out claims that were each to yield more than a million dollars, and this tributary of a tributary came to be called "Eldorado."

Big Alex McDonald of Nova Scotia hit a pay streak 40 feet wide, "and a man could, and did, pan $5,000 dollars from it in a single day."[14] And, finally, Clarence Berry's claim had so much gold on it that when his wife needed spending money "she merely walked to the dump and with a sharp stick, smashed apart the frozen clods and pulled out the nuggets."[15]

What was probably the world's last great gold rush happened in 1979 in a green, jungle wilderness called Amazonia. This gold strike affected the countries of Ecuador, Peru, Colombia, Guyana, Venezuela, Bolivia, and French Guiana, as well as Brazil; "but it is only in Brazil that the rush has reached a level comparable to the great gold rushes of the last century." Here the number of miners would make Dawson look like a ghost town; "at any one time numbers run into the hundreds of thousands and they produce scores of tons of gold worth, in the late 1980's, over one billion annually."[16]

The reason so many miners are needed is because most of them are walking coffins with a life expectancy less than that of a new TV show. Believe it or not the miners sprinkle liquid mercury (as toxic as toxic can be) behind the riffles in a sluice box. The mercury amalgamates with the gold, making it heavier and thus more likely to be retained in the bottom of the sluice box.

"At the end of the day the amalgam is tipped into a piece of finely woven cloth, which is often merely the T-shirt the [miner] is wearing, and twisted around hard to expel any further excess mercury. The amalgam is then heated with a butane torch: the mercury evaporates away in a toxic white gas, the gold glows red hot for a few seconds," then cools to its normal yellow color. The miners commonly prepare this death tonic right where they cook their food.[17]

The world's first gold rush occurred in ancient times in the land of Colchis, bordering the Euxine (the Black Sea). Here placer miners extracted gold by shoveling white gravel into sluice boxes made out of hollow trees. A lining of sheepskins entangled the gold particles (today' miners use astroturf); the coarser gold was shaken out, but the fine gold adhered to the wet wool, which was hung on trees to dry. Once dry the gold was beaten out of the fleeces like dust from a carpet. A tale of this gold strike reached Greece, and Jason and his Argonauts set sail in the good ship *Argo* to seek the golden fleece.

And what of the world's oldest and richest placer deposits? These occur in the Witwatersrand Basin in South Africa and are believed to be more than 2.2 billion years old. The gold is found in a hard, Precambrian sedimentary rock called conglomerate (solidified gravel). Mined since 1886, these deposits have dominated world gold production for most of the twentieth century; in fact, they have produced almost half of all the gold mined since 1900 (over 36,000 tons). However, the mines are now so deep (12,000 feet, or 3,658 km) that it is questionable how much longer they can safely be worked. The source, richness (0.2–0.7 ounces per ton), and amount of the gold have been the subjects of fierce debate for more than a century. Today there are still no definitive answers as to why so much gold came to be concentrated in a place called Witwatersrand.

THROUGH THE CRYSTAL CAGE

Diamonds, souls shinning through crystal cages, are not only cherished and flashy but also the hardest natural substance found on earth. Graphite, the "lead" of pencils, is soft, greasy, and as black as burnable coal. Diamonds, graphite, and coal, about as physically diverse as any three substances can be, yet each is related to the other in much the same way families are linked through genes. These three—beautiful diamond, soft graphite, and glowing coal—are all composed of the element carbon. One element—three very different offspring.

It turns out that diamonds are hardboiled under very high temperatures (2,552–3,272° F [1,400-1800° C]) and pressures deep within the earth's crust, down 93–186 miles (150–300 km), down where the forces that move continents toil, and where the music of the spheres plays.

Graphite and coal are stewed and simmered at much shallower crustal levels, up in the realm of volcanoes and earthquakes. This slow roasting turns carbon black, flat, and soft; its nature and structure are about as far removed from those of diamond as Haiti is from winter. Strangers they be and when they meet, well:

"The Coal and The Diamond"

A coal was hid beneath the grate,
(Tis often modest merit's fate;)
'Twas small and so perhaps forgotten;

Whilst in the room and near of size,
In a fine basket lined with cotton,
In pomp and state a diamond lies.
So, little gentleman in black,
The brilliant spark in anger cried,
I hear in philosophic clack,
Our families are close allied;
But know the splendor of my hue,
Excelled by nothing in existence,
Should teach such little folks as you
To keep a more respectful distance.
At these reflections on his name,
The coal soon reddened to a flame:
Of his own real use aware,
He only answered with a sneer;
I scorn your taunts, good Bishop Blaze,
And envy not your charms divine;
For know I boast a double praise,
As I can warm as well as shine.

—Anonymous

However, dear coal, remember that diamonds are forever and the symbol of eternal love—or so diamond sellers would have you believe.

In geology what goes around comes around, and majestic diamond, on entering the shallow realm of humble coal and lowly graphite, becomes unstable and starts to fall apart. At these shallow depths a diamond's crystal structure begins to deteriorate. So, with enough time, good Bishop Blaze changes to graphite. In fact, some kimberlites (diamond-bearing rocks) contain perfectly shaped diamond crystals that are now nothing but pencil lead. Despite Madison Avenue, diamonds aren't forever. It may take several million years, but one day many of Elizabeth Taylor's fine jewels will be the lead in someone's pencil.

Diamond, which is derived from the Greek word *adama*, meaning unconquerable, is a very rare mineral. Since its discovery, around 480 B.C. in India, it has been highly prized as a curiosity object, considered a living spirit (Plato thought this), and believed to keep away phantoms, bad luck, and lunacy, and to make the sleep of the wearer free from nightmares. Diamonds were also believed to cure the sick and heal the sinner. In both cases the "helped" person had to hold the diamond in his or her mouth. The sick person did this while fasting (and how long do you think this went on for?), whereas the morally slothful sucked on it as they confessed their sins (that way only God could understand them).

To tell a true diamond from an imposter, Pliny the Elder wrote, "These stones are tested upon the anvil, and will resist the blow to such an extent as to make the iron rebound and the very anvil split asunder."[18] Of course, like some diamond ads, this is not true, and over the years many fine crystals were broken and destroyed because of this belief; diamonds, when struck a hard, direct blow, will splinter or powder. However, such a belief was a real boon for dishonest merchants and dealers. Before they would buy diamonds from a seller they would test them by hitting the stones with a hammer. Of course, the diamonds would splinter and the disgusted seller would toss the rest of the "worthless" stones away, believing he/or she not only had tried to sell fake stones but was also a fool for believing them to be real. The dastardly dealer would then scoop up the abandoned diamonds and make off with the loot.

Though diamond can be pulverized with a sharp blow, it is extremely difficult to compress or deform. It is said that a crystal placed between the jaws of a steel vice will actually penetrate the hardened steel as the vice is tightened. However true this may be I certainly wouldn't try it with the family jewels.

Generally, when you think of diamonds you imagine jewels, gems, white ice, crystal fire, and the glitter and sparkle of tiaras and pendants at operas and the Academy Awards, but diamonds are much more than that. They are prized both as gemstones and as the most important abrasive known to humans (right after taxes and lawyers). The industrial demand for diamonds can be attributed to the mineral's hardness, its great thermal conductivity, and its high resistance to compression.

Diamonds, both gemstones and industrial stones, are sold by the carat. The word *carat* comes direct to us from the Middle Ages and the Middle East, where it was the weight of one dried carob seed—about $\frac{1}{2000}$ of a pound (imagine the crafty dealers trying to figure out ways to increase that weight with the old lead in the carob seed trick). Today we are more mathematical, so the carat is defined as $\frac{1}{5}$ of a gram (0.006 troy ounces). The term *point* is also used in the diamond business and means $\frac{1}{100}$ of a carat, so a 10-point stone is pretty small.

Based on their occurrence in nature, diamonds are divided into four classes: (1) gemstones, (2) industrial stones, (3) bort, and (4) carbonado, or carbons.

Bort is the name given to microcrystalline diamond masses. When these are pulverized they can be used as polishing grit. Carbons are those diamonds that are at various stages of transition from diamond to graphite. These stones can be used in industry if they haven't gone too far down the graphite road.

Gemstones, one and all, have four common characteristics: they are beautiful, durable, rare, and expensive. In diamonds this means largest size, fewest inclusions, most perfect crystals, and overall best quality. Dispersed nitrogen in the crystal structure of a diamond can give it a yellow, green, or blue cast, thus adding to its rarity, beauty, and value. Raw gemstones are worth anywhere from $100 to $5,000 per carat depending on their quality. Generally quality is defined as "whatever the official appraiser says it is!" Eighty percent of all set gems are used in wedding and engagement rings. The world's largest consumers are the United States (about 45%) and Japan (about 25%).

Throughout history there have been diamonds and then there have been DIAMONDS, the big, the beautiful, and the famous.

The Koh-I-Noor (mountain of light) was worn by rajahs as long ago as A.D. 1300. This 186-carat stone was eventually ripped off by Mogul emperors and used as one of the eyes of the peacock in the famous "peacock throne." Eventually, the East India Company seized it for debt payment and gave it to Queen Victoria. It is now one of the crown jewels and rests in the Tower of London.

The 410-carat Regent diamond, also found in India, was originally called the Pitt diamond. It was sold to the Regent of France (thus its current name) and was worn by Marie Antoinette. A few years and a republic later, Napoleon had the diamond set in the hilt of the sword he used to crown himself emperor.

The smaller but more deadly Hope diamond is a 44-carat blue diamond that now lives at the Smithsonian Institution. The previous owner, while in possession of the "cursed" stones lost a child to an accident, saw her family break up, lost her fortune, and committed suicide. If diamonds are made down where celestial music plays, then maybe they are meant to be worn only by gods and goddesses—Great Zeus, Holy Aphrodite, and Elizabeth Taylor.

The largest diamond found in the United States is a 40-carat stone, called the Uncle Sam, from a mine in Murfreesboro, Arkansas (this site is now a park where you can go and dig for diamonds). The biggest of the big, found in 1905, was called the Star of Africa (better known as the Cullinan). It weighed in at 3,026 carats (1.5 lbs, or 0.68 kg)) and was cut into 105 separate stones. The two largest of these are still the biggest diamonds in the world at 530 and 312 carats. They were both given to King Edward VII and are now part of the Crown Jewels.

Until the late 1920s gems were the only common use for diamonds, and subquality stones were considered a waste product. Extensive research into this "waste material" sparked new uses for non-gem-quality diamonds. This "experiment" was so successful that by 1934 over 70% of the market was for industrial stones. This trend has continued and today about 80% of all diamonds are used for industrial purposes. Natural industrial diamonds sell for $1 to $4 per carat.

Industrial diamonds can be grouped into two classes depending on size and use. Rough stones consist of fine-grained grits used for polishing, grinding, and sawing. Then there are shaped stones (also called tool and die stones) in which the orientation of the diamond's crystal structure is important. Since a diamond abrades or wears down much faster in a direction perpendicular to its crystal face, the stones are set so the crystal face is parallel to the work surface. Uses for shaped stones include scalpels and diamond drill bits.

Diamonds have been made synthetically since 1955. Today about 60% of industrial stones (around 150 million carats) are produced synthetically each year. General Electric uses a process that involves dissolving carbon in molten silicate metal. This material is then heated to 5,400° F (2,000° C) and put under a pressure of 1 million pounds per square inch (the Michelin man would be the size of a sand flea). Using this technique, grit-size stones can be mass produced in 2- to 3-minute cycles.

Geologically (or *au naturel*) diamonds are found in only two types of economic concentrations (generally meaning 20 carats of gem-quality diamonds for every 100 tons of rock): in rocks called kimberlites (and a closely related rock called laprolite) and, when these get deeply weathered, in placer concentrations (stream and beach deposits).

Kimberlites, named for Kimberley, South Africa, where diamond-bearing rocks were first identified and mined, are believed to form at crustal depths of more than 100–125 miles (161–200 km) and temperatures of 2,012–2,732° F (1,100–1,500° C. Molten kimberlite material is then emplaced rapidly (upward speeds of 6–20 miles per hour (10–30 kmph) into shallow crustal environments (emplacement has to be this rapid to keep diamonds from doing their graphite thing). Emplacement occurs along deep (100 mile or 161 km) faults or breaks and only in what geologists call "stable continental areas," places where the earth's crust has done nothing much for at least 1,500 million years. This being the case, kimberlites usually hang out with Precambrian shields. They occur in South Africa, India, Australia, Canada, and in the United States, in Wyoming, Colorado, and Michigan. Individual kimberlites are pipe-shaped bodies that vary from a few feet to a couple of miles in diameter. Diamonds may or may not be associated with them. Of the 100 kimberlite pipes in the area around Kimberley, South Africa, only 20 have diamonds; of all the kimberlite bodies known around the world only 1,000 contain diamonds, and of these, only 60 are economic concentrations.

The diamonds found within kimberlites are actually a lot older than the kimberlite within which they are found (now that's pretty good geologic doublespeak). This has been determined by age dating of inclusions that occur within the diamonds. Based on age dating, the ages of most diamonds are either 3.3 billion years old (before life), 1,150 million years old (before the Volvo), or 950 million years old (before CD players), whereas the associated kimberlites give ages of only 100 to 1,100 million years. This means that kimberlites are nothing more than taxi cabs to the surface for diamonds. These age dates suggest that diamonds have been forming throughout much of the earth's history, and may be a natural and very common mineral deep, deep down in the heart of the crust. They are material that has been stolen from what geologists call the "mantle" by those fleeing felons—the Kimberlite band. It has recently been suggested that asteroids, those dark, rocky lumps that sit between Mars and Jupiter, are simply stuffed full of diamonds, and all someone has to do is figure out a way to get out there and mine them. Ah, finally, a cost-effective project.

Prelude to Chapter 7

The whole expanse of a glacier. A cold vastness of wind, snow, and ice. Ice that is on the move, constantly creaking and scraping, cracking and snapping, as it slowly flows outward to cover the land.

Hear the slow, harmonic progression of chords in the French horns as the cold comes, ice forms, thickens, and begins to flow. Introduced into this background comes a long melody by the cello. Sometimes this melody is lyrical, as the ice slowly flows across the land; at other times it is fierce and improvisatory, as the ice cracks and creaks, or snaps completely off at its front to form icebergs.

CHAPTER 7

Winter's Breath

The two of them had been alone for the last 3 months. Dressed in their warmest robes, they stood hand in hand staring over the cold, wet plain. Where the level ground had once risen into low, rolling hills there was now only It—the eater of hills, the destroyer of rivers and fish, a great walking wall of blue ice whose roar caused the wind to run and the earth to shake. Heaven to stars had it roared last night. Long, sorrowful creaks and groans, air-whooshing pops, and loud, crackling snaps like trees breaking in the wind. It had been impossible to sleep, and now, in the gray of the morning mist, they saw why. During the night It had slid off the hills and onto the plain.

They believed now. They were lost like the rest of their clan. There was no end, no rest; only cold and fear and flight. Like the horned ones and the ones with the thundering feet they would again run before It. Squeezing their hands together, in a moment of love and determination, they turned and entered the smoky cave. The man began gathering up their tools, weapons, and pots, while the woman added the last of the fleeing creatures to the wall drawing. These were creatures that walked on two legs, a man and a women following the herds, their backs to a great, dark shape.

They needn't have left. That winter brought less snow and milder temperatures than the previous 10 decades, and the following summer was so warm the great ice sheet bled. This was the start of the meltdown, and it went on year after decade after century after millennium. It, like a wounded animal, crawled slowly north, its margin searching for the safety of its polar den.

The glacier melted, but not peacefully. It left great footprints all over the land that would remain for tens of thousands of years. Deep valleys and basins had been formed and were filled with the glacier's own blood, to make small and great lakes. Tall hills had been bulldozed flatter than one of Aunt Jemima's pancakes, and mighty rivers had been forced to turn tail and run in opposite directions. A giant landfill of rocks and sediment that geologists call till, outwash, and erratics (the Three Magi of a glacier) was strewn from one end of the continent to the other.

The ice took 8,000 to 10,000 years to leave North America, but has it gone for good? In examining the earth's geologic history there is strong evidence to suggest that ice ages have been with us for more than three billion years. True, they may be rare as geologic phenomena go, but give or take a couple of hundred million years and you can just about guarantee one. So the real question is not whether we will again be visited by ice, but whether we humans will be around to fight or to flee. Over 9,000 years have passed since a man and a women stood alone on a rocky ledge. Yet we can still reach back through the millennium and share their fear as well as their determination to survive:

> There is something of our glacial past
> Buried deep inside each of us,
> A primitive knowing,
> For why else do we welcome the April rain,
> And find such joy in the first wild flower of spring.

But with the dust of ages has come a new understanding of our rhythmic planet.

> The earth spins on,
> Sending continents sliding

across its moving face,
Some south, some north,
North toward the poles,
North to the snow and cold,
Ever north to give birth
To a new age of ice.

It's hard to imagine what effects an ice age would have on climate (ski resorts open all year round), sea level (our great port cities would be high and dry and some 50 miles from the edge of the sea), and lifestyle (no more concerts in the park). Fortunately, we live in a geologic time when the ice has retreated to the highest mountains and the highest latitudes. Today ice covers only 10% of the land area, but surface features formed by glaciers are found on 75% of the earth's walkable land. Through the eons of geologic time, each and every continent has been touched by the joy of glacial winter.

Glaciers are really nothing more than great lumps of ice on land that have the ability to move under their own weight (just like poured pancake batter in the middle of a griddle). The most extensive glaciers in the United States reside in Alaska, where they occupy 3% of the state's land. Other glacial states include Washington, Montana, California, Idaho, Colorado, Oregon, and Wyoming. All of the ice lumps in these states belong to the fraternal order of alpine, or mountain, glaciers. There are over 200,000 members worldwide, and these are mostly local, stay-at-home glaciers living in mountains and mountain valleys.

The other order of glaciers is bigger, bolder, and better. These are international organizations of ice called continental glaciers, or ice sheets. They cover entire continents, reach thicknesses of a mile or more, and are so heavy they cause the earth's surface to sag downward for thousands of feet. Fortunately for us, membership in this organization is so exclusive that there are currently only two members: the Greenland and the Antarctic ice sheets. The Antarctic glacier is so large it could easily cover the 48 contiguous states to a depth of 1.5 miles (2.4 km). Not only are they big, but ice sheets are also old, so old they make Methuselah look like a spring chicken.

An alpine, or mountain, glacier. This Alaskan glacier is one of 200,000 mountain glaciers found around the world. (Photo courtesy of C. L. Matsch)

Continental glacier, or ice sheet, which can cover entire continents, burying the land beneath hundreds of feet of ice. (Photo courtesy of C. L. Matsch)

Some of the ice in the Greenland glacier is more than 250,000 years old. Now here is ice that deserves to melt in a very rare and very old single-malt scotch whisky.

Glacial ice is for drinks (in Japan you can get glacial ice cubes—for a price), for extra-pure mineral water (only in Iceland, thank you), and, in the future, most likely just for water. Over 75% of all the world's fresh water is stored as ice in glaciers, and 99% of all this frozen liquid is found in the Greenland and Antarctic ice sheets.

To get an equal amount of water it would have to rain steadily over the entire planet for 60 years (and I thought a winter in Seattle was bad).

FRANKENSTEIN'S MONSTER

Glaciers are the Dagwood Bumsteads of geologic phenomena. They eat huge quantities of food (snow) and in doing so gain considerable weight. However, unlike their human counterparts, where a large weight gain turns muscle into soft, sagging rolls of fat, a glacier's increased bulk transforms soft, fluffy snow into hard, lean ice.

Snow to ice takes time, and some very special earth effects. Plant a lot of snow, freeze well, and anywhere from 10 to 3,500 years later, depending on the rate of snow accumulation and climate, you can harvest a ton of ice (unless you live in Houghton, Michigan, where the whole process takes about four months. At least that's the perception of the snowbound residents!). The special earth effects come in the form of a large landmass occupying a polar region (for a continental glacier), or land with really tall mountains (for alpine glaciers), and a climate cold enough to be a good host to snow for much of the year.

These are the physical conditions that make a glacier fit and healthy. To make it get up and roam the land, the amount of falling snow has to exceed the amount of melting snow for longer than it takes Congress to agree on something.

As the accumulating snow slowly gets deeper and heavier, it scrunches the zillions of snowflakes tightly together, forcing all the air out of the collapsed stars, and bringing on a midlife crisis. The once young, delicate, light flakes are transformed into middle-aged crud, a dense, granular, crusty snow called corn snow, or firn (this is the same hard, dirty, dog-marked stuff found along city streets at winter's end). Middle-aged corn snow, with a cup more of compression and a couple of gallons of time, is reborn in old age as clear, crystalline ice.

Over time, as snow continues to fall and accumulate, changing slowly from firn to ice, a point of no return is reached. One day the ice, like Frankenstein's monster, simply gets up and walks. Ice, as either cubes or glaciers, begins life as brittle as a fine crystal glass. However, when it has enough weight on top of it (about 165–200 feet of snow [50–61 m]) it begins to lose its brittleness and starts to take on the character of pancake batter. Take a big ladle full of well-mixed pancake batter and slowly pour it onto a flat tabletop, or into a pan. As you watch, the batter slowly flows or oozes

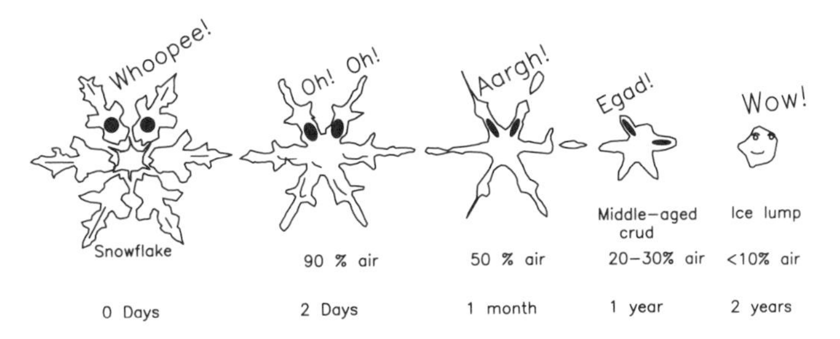

The growing season: the slow transformation of snow to ice. (After Bader et al., 1939)

outward in all directions. Like pancake batter, glacial ice oozes (or flows) outward and downward. Geologists call this "plastic flow" (which sounds to me like a face-lift gone bad). Glaciers that move by plastic flow are thick enough to travel under their own weight, similar to the outward movement of the pancake batter.

Pancake batter glaciers "flowing" out of valleys in Alaska. (Photo courtesey of C. L. Matsch)

Ice that moves by being plastic doesn't move all its parts at the same speed. The base of the glacier will drag over the ground, thus moving slower than the ice above it. The very upper part of a glacier has little or no weight on it, and will remain as brittle as the day it was formed. This ice is carried along piggyback style by the flowing ice below it. Any change in the rate of flow, or in the direction of flow (which can be caused by topographic irregularities, such as Charlie's barn), leads to the brittle fracturing or pulling apart of the top portion of the glacier, and the formation of deep crevasses.

Flowing ice travels at one of two speeds—slow and slower. At the slowest speed it's about as fast as a hound dog on a hot day in Georgia; open the throttle all the way and we're talking snails racing up aquarium walls (about 65 feet a year [20 m]). However, there are special occasions in the life of a glacier when it will surge or gallop forward at tens of feet per day. The Hubbard Glacier, in southern Alaska, is one such surger, moving up to 30 feet (9 m) a day. The Variegated Glacier northwest of Juneau could be called the roadrunner of ice—it can gallop as much as 200 feet (61 m) per day. Galloping glaciers are usually thin and located in places where the summer temperatures are warm enough to cause the ice to melt, forming a thin sheet of water at the glacier's base over which it then slides, similar to children slipping and sliding down a wet, yellow plastic sheet called a "wet banana."

It was a roadrunner that Mark Twain was hoping for when he wrote about making camp on the surface of a glacier. He was expecting a fast, easy ride right to the valley bottom. With lawn chairs out and cigars lit, Twain's party relaxed and waited for the scenery to fly by.[1] Unfortunately, they had chosen a run-of-the-mill glacier that turned out to be slower than molasses in January. If they had had the patience to wait, they would have reached their destination in about 1,000 years (and that's a lot of cigars).

As snow continues to fall and the glacier gains weight, ice flows outward and downward. Somewhere, sooner or later, the front of the advancing ice meets its end. This can come about when the glacier runs into a large body of water, and begins to split up as icebergs. Icebergs can be small (one-penguin rafts), or absolutely gigantic. The biggest iceberg ever observed broke away from the Ross Ice Shelf in Antarctica in 1987. This was no burg; it was a great, rip-roaring country with a length of 100 miles (161 km) and an area of 2,412 square miles (6,250 km^2), and it contained enough fresh water to supply the city of Los Angeles for 675 years!

The other fate awaiting for the leading edge of a glacier is warm air. When ice works its way to places where the temperature is warm enough, the ice will melt faster than the glacier can resupply it, and you can hear those church bells toll.

An advancing glacier is said to be one in which snow accumulates faster than it melts. Such a glacier grows thick around the middle, and extends farther and farther into new territory. When the rate of loss through melting (called ablation) is greater than the rate of accumulation, the glacier will beat a hasty retreat back to where it feels secure, and we thus have a retreating glacier.

Over many years, if the climate of an area remains relatively stable, a glacier will become lazy and complacent. The amount of snow accumulation in the colder regions, will be about equal to the rate of melting in the warmer areas, so the front of the glacier remains in about the same place.

As the seasons turn, glaciers do their thing (melting and shrinking, growing and advancing) under what geologists call a strict budget. Unlike governments, if a glacier's budget runs in the red for too many years, it's history. But some glaciers spend ice and accumulate snow at such rapid rates it would take all of H & R Block's accountants to keep track of the gains and losses. The Franz Josef Glacier in the southern alps of New Zealand is one of these. It has an enormous snow accumulation of between 350 and 500 inches a year (889–1,270 cm). With so much snow you would think it would be growing and advancing at a very high rate. Wrong. This glacier advances alright, but it does so straight down a mountain into a lush, green rainforest. Here the ice undergoes instant meltdown and spurts water like a broken fire hydrant. The combination of a large snowfall in the mountains, and an ice front in a warm rainforest, causes the ice to advance and retreat anywhere from a quarter of a mile to more than a mile per year (0.4–1.6 km).

On the other side of the glacial ledger are ice sheets, which are slow to respond to climate changes. It takes many years to get them to advance or retreat any significant distance. The Antarctic ice sheet gets only a few inches of snow a year, and ice loss takes place mostly by the "calving" of icebergs (the breaking away of a mass of ice from a glacier) into the sea (wouldn't this make a great "Far Side" cartoon?). There is not much accumulation or melting, so the ice stays pretty much put. It would take years of greatly increased snowfall, or of increased temperatures, to have any effect at one end or the other.

SANDPAPER MASSAGES

When they flow, slide, surge, or melt, glaciers are working. They work 24 hours a day, 365 days a year without a vacation or fringe benefits. They work hard to polish, move, and grind rocks; to dig valleys and basins the likes of which the Corps of Engineers only dreams of; to bury the land under a thick blanket of debris; and, in more human terms, to mold and fashion mountains as monuments to their power and glory.

The landforms created by continental ice sheets are not as spectacular as those formed by alpine glaciers. Alpine glaciers continually grind away, giving us grand panoramas of U-shaped valleys, steep canyon walls, jagged mountain peaks, sharp ridges, hanging valleys, and waterfalls. Continental glaciers simply overwhelm the land. They bury it, creating a topography that is as close to a pancake as an oboe is to an English horn.

Glaciers scrape, scour, pluck, rip, grind, and crush; in short, they terrorize rocks of all sizes and shapes. Rocks are picked up from the ground the glacier travels over, and plucked from the sides of valleys and basins. They give the glacier its scouring power, which geologists call abrasion.

Glacial ice, all by itself, is too soft to erode and polish the rocks it flows over. However, any loose rocks that are picked up along the way become frozen solid into the ice. These are then dragged

Glacial striations, or grooves, on a rock surface in Utah. (Photo courtesy of C. L. Matsch)

along by the glacier, turning the smooth, flowing ice into coarse sandpaper. This glacial sandpaper, as it slides along the ground, can gouge fine, parallel scratches in bedrock, which are called striations, or grooves.

The continued scouring and grinding action of a glacier slowly breaks the rocks into smaller and smaller pieces, so that they eventually become a powdery material called "rock flour." Most glaciers produce so much rock flour that meltwater leaving the ice has a milky white appearance similar to skim milk.

Glaciers also erode and pick up rocks by a process called plucking, which occurs where meltwater is able to seep into broken and cracked rocks at the base of the ice, and freeze solid. This frozen water attaches the broken rock firmly to the bottom of the glacier; as the ice flows forward, the stuck rock is pulled out of the ground like a bad tooth attached by a string to the back of a train. Over long periods of ice advance, this process makes bowl-like depressions called cirques, which are commonly found at the head of alpine glaciers. When they are filled by water, cirques form small lakes

A smooth, curved rockform, or whaleback, caused by glacial plucking or quarrying. (Photo courtesy of C. L. Matsch)

called tarns. The basins that hold the Great Lakes were deepened in much the same manner, and then filled by water after the last continental ice sheet wasted away.

Plucking, or quarrying, also leads to the formation of smooth, elongate, curved rock forms called whalebacks. These formations have been given this name because they resemble the backs of whales headed out to sea. These features occur solo or in schools, and the upper surfaces of whalebacks are commonly striated or grooved.

GARBAGE TRUCKS OF THE PLEISTOCENE

Glaciers are like garbage trucks. Have you ever been startled awake at 7 a.m. by the clanking and banging of garbage cans? If you are curious enough to get up and stumble to the window, you will see trash cans being emptied into a gaping, steel-jawed monster. With one push of a button the jaws crunch closed, and the garbage vanishes.

Glaciers, like garbage trucks, gobble up earth debris. They eat rocks, soil, trees, sediment, freeways, and anything else that's too slow to get out of their way. All the stuff they eat resides somewhere inside the icy giant, and when the glacier melts, as they all do, the resulting mess provides a special Kodak moment. It's a mess far worse than that left by a theater full of kids with boxes of sticky candy, buttery popcorn, and cups of ice-filled pop. It's a huge landfill of sediment and rock tens to hundreds of feet thick that levels out the countryside to make mile after mile of rolling, rocky pasture land. Minnesota is one of the last ice age's five-star pastures. Here, glacial garbage covers 98% of the bedrock and, in places, is almost 1,000 feet (304 m) deep.

A glacier disposes of its garbage in one of two ways. First, it simply leaves without picking it up. The glacier just melts away and viola, there's a house-size boulder sitting on top of a cliff or in the middle of a plain, such as the 20-ton boulder of dark green basalt that sits out on the floodplain of the Chippewa River in southwestern Minnesota. Years ago local residents realized that this giant was completely different from the pink granite that formed the local bedrock. Knowing that this was a stranger from a strange, strange land, they came to believe it was a visitor from outer space and so named it the Montevideo Meteorite.[2]

The other way a glacier dumps its garbage is by the sweat of its own blood. Water from the melting ice carries tons of sediment and rocks away from the dying beast, and deposits it across the countryside.

Both of these disposal methods create a huge variety of glacial landforms and landfills.

Glacial erratics are large boulders carried by the ice far from their birthplace. These are the rocks from outer space, the "wonder how that got there" rocks, boulders the size of houses and cars sitting on hills, in swamps and forests, and in open fields. Early Americans recognized them as something special and so they

Poorly sorted and unlayered glacial till. (Photo by C. L. Matsch)

An elongate, streamlined hill formed by a glacier is called a drumlin. (Photo courtesy of C. L. Matsch)

painted them different colors, and put designs on them in honor of their strangeness and size.

Till is a wondrous word. It is Scottish for stiff, rocky ground. You hear Scotsmen talking about till all the time. "Aye Laddie, gonna doone to the till for just a wee bit." Till, when translated into the Queen's English, actually means "still," and the Scotsman's "stiff, rocky ground" refers to the shape he's in after a wee bit of time spent there.

In glacial terms till refers to sediment and rocks deposited directly by the melting ice. Till is not layered, or bedded, and is said to be poorly sorted (which means large and small rocks occur side by side). Lithified till is called tillite. Till, in North America, should invoke visions of rolling, rocky pastures enclosed by stone fences.

Before 1837, the year a Swiss zoologist by the name of Louis Agassiz proposed the idea of ice ages, some parts of glacial till were called Noah's barnyard. This was to organic material brought to

Water-deposited glacial sediments are generally called outwash. Such deposits are layered and sorted. (Photo courtesy of C. L. Matsch)

the surface while digging water wells in till-covered ground. The material consisted of plant and soil remains mixed with glacial sediment. Early theories on the origin of till and glacial erratics placed them as materials left from the great biblical flood. The strange house-size boulders were believed to have been floated in on icebergs, and the organic material was the remains of Noah's floating stable.[3]

Landforms constructed by glacial till are called moraines. This is another Scottish term derived from the old Laird's saying, "If ya can see across the Loch, laddie, it means we're ta have more rain; if ya cannot it means its already raining." Glacially speaking there are three different types of moraines. Ground moraines are broad blankets of ice-deposited till. These can be tens of miles wide and hundreds of feet thick. They fill in depressions and smooth out landscapes. An end moraine is a ridge of till that forms at the end (front) of the glacier. These occur only if the front of the glacier has

A winding ridge of glacial sediment formed by a meltwater stream is called an esker. (Photo courtesy of C. L. Matsch)

been stuck in the same place for a long period of time. At such stagnant ice fronts the glacier continues to bring in rocky debris which, as the ice melts, is deposited to form ridges tens to hundreds of feet high and miles long. Morainic deposits that are trampled over by a later ice advance become smeared out into elongate, streamlined hills called drumlins. Terminal moraines are ridges of till that mark the spot of the glacier's furthest advance to the south.

Melting glaciers not only deposit sediments and rocks but also let loose tens of thousands of gallons of water that redistribute and deposit sediments. Water-deposited glacial sediments are called outwash, and, unlike till, this material is layered, or bedded, and much better sorted (the material occurring together is about the same grain size). Landforms associated with outwash deposits include:

1. Eskers—winding ridges of sediment formed by meltwater streams flowing within and at the base of a melting glacier.

This end moraine has damned a small glacial valley. Water from the retreating glacier has formed a small glacial lake behind the moraine. (Photo courtesy of C. L. Matsch)

2. Kettles (not the drums)—form where a glacier melts back so fast it leaves large blocks of ice stuck in the glacial drift. When these blocks melt they leave behind a hole that is often filled by water. These depressions can be up to 6 miles wide (10 km) but are generally less than 50 feet deep (15 m).
3. Outwash plains—similar to ground moraines and are broad sheets of outwash sediment.
4. Kames—I kame, I melted, I deposited. These are steep-sided, conical hills composed of sand and gravel. They form when sediment collects in openings or cracks within stagnant ice fronts.

End moraines may also form dams within, or at the end of, glacial valleys or depressions. Like modern dams, these trap water behind them, leading to the formation of glacial lakes. Typically, sediment (mostly rock flour) carried by meltwater pours into such lakes during the summer months (melting season), forming layers of relatively coarse sediment; finer-grained sediment accumulates or settles out of the water during the winter season, forming finer-grained layers. A set of these winter–summer layers is called a varve, and by counting these, just like tree rings are counted, an estimate of the number of years it took to form the total thickness of sediments can be determined.

Glacial lakes come in all sizes, from the neighborhood pond to the Great Lakes. The largest of the Pleistocene glacial lakes was called Lake Agassiz. At any one time this body of fresh water covered 50,180 square miles (130,000 km^2) to an average depth of 400 feet (122 m). It covered large parts of Manitoba, Ontario, Minnesota, and Alberta. As the Pleistocene ice sheet melted away, Lake Agassiz shrank, leaving several separated remnants, such as Lake Winnipeg, Lake Manitoba, and the Red Lakes of Minnesota.

Loess is fine-grained sediment picked up by the wind from glacial outwash surfaces and deposited elsewhere. Loess, if thick enough, forms rich soil that holds moisture well and is easy to cultivate. Loess is also easy to dig into, so insects, rodents, and humans (a royal triumvirate) bore, tunnel, and channel into it. People in inland China have, for generations, lived in loess caverns. These dwellings are easy to maintain, warm, and well-insulated. However, they do have one serious flaw—they are situated in earthquake country. When the ground ripples, the houses of loess tend to shake apart.

THE GRINCH WHO STOLE SUMMER

In Norse mythology the chilly breath of the frost giants blew over the land and a giant wolf swallowed the sun. The earth grew cold and the long winter came. The rivers and inlets froze solid and the long ships scraped and groaned against the ice like animals caught in a hunter's trap. The forests were cut and burned for warmth, and food became a luxury. The year was 1350 and the Little Ice Age had begun. It lasted about 500 years.

During the Little Ice Age, snow lines (the line above which snow remains on the ground all year around) moved 650–1,400 feet (198–426 m) closer to sea level. Across Europe crops frequently failed, famine was widespread, and on several occasions the Baltic Sea froze solid, and people could skate, slide, and ice fish from Germany and Poland to Sweden.

If a "little ice age" refers to cold, snow, and starvation, what does a full blown, continent-busting ice age refer to? Not a cold war or the way certain beer is brewed, but to a period of time when the earth is cold enough to allow glaciers more than 386,000 square miles (1 million km^2) in size, with ice more than 1 mile thick, to form on nonpolar continents. During the last ice age, glaciers covered all of Canada, as well as the middle and northeastern parts of the United States. Over 6.2 million square miles (16 million km^2) of ice sat upon North America and, worldwide, more than one third of all the land was covered by frozen water.

With such tremendous volumes of water locked up as ice, sea levels fell 300–400 feet (91–122 m). In those days there was no need for a tunnel from France to England, and you would have to drive 50–60 miles (80–97 km) out of Cape Cod just to reach the ocean at high tide.

When you think about it, ice ages are nothing but trouble unless you're a polar bear, penguin, or ice worm. Speaking of ice worms, I always thought they were a figment of the Alaskan tourist bureau's imagination. They sell these postcards with pictures of piled-up spaghetti on a bunch of snow and call them ice worms. Who would believe that the little beasties actually exist? Well they do. They reside in corn snow and pools of meltwater near, or at the top of, glaciers. About an inch long (2.54 cm) and a quarter inch thick (6.3 mm), they crawl and wiggle through the new-formed ice, feasting on algae and airborne spores. They are a well-kept secret, probably because speckled trout adore them.

For the rest of the living world an ice age would mean mass population shifts (it would be like Disney World at Easter), rivers changing course, some lakes destroyed and new ones born, and large drops in sea level (lots of additional room for new housing developments, but don't ask about flood insurance), making your house at the seaside prime desert property (left high and dry).

Obviously, we have a lot to lose in an ice age, so it makes sense to try to find out what causes these frigid periods of time and what we can do to help the earth avoid another one. Unfortunately, ideas, theories, and guesses on the reasons for ice ages are as common as bad TV commercials.

To get an idea on the state of current thinking, I attended (without invitation) an international meeting of the most distinguished glaciologists and climatologists the world has ever seen. I arrived at the great hall after the session had started, so I quietly slipped in through the rear doors. Silent as an ice worm, I found a seat in the back and sat to listen to the words of wisdom and logic pouring forth.

"Ice ages are due to a change in the tilt of the earth's axis." So lectured a well-dressed young man, who simply couldn't leave his oxford tie alone. He also had the unfortunate habit of standing on his tiptoes and rocking back and forth as he spoke. The podium could have been the deck of a sailing ship and he its master at the wheel. I wondered how many others in the hall had that queasy feeling in their stomachs that marks the onset of sea sickness. "Such a change," he said, swaying backward, "would cause a change in the distribution of sunlight that reaches the earth," and forward he swept. "If the poles were tilted farther away from the sun," while back he went, "the polar regions would become colder," and forward he swayed. "And, if a large land mass occupied that region," as back he tilted, "a great ice sheet would begin to form," and forward he bent. "Now, I think—"

"Humbug and bogtrotters," rudely yelled a bearded gentleman in a dark blue jacket that had dribbles of egg yolk down the front. "Absolute rubbish," he roared. "Ice ages are due purely and simply to volcanic ash and gas thrown high into the atmosphere by great eruptive events. This material is carried around the globe, forming a blanket that blocks out incoming sunlight, causing a decrease in temperature. If we examine weather records after great eruptions like Krakatoa, the associated temperature drop would be more than sufficient to initiate an ice age."

"Poppycock," cried a small, gray-haired lady in a bright blue dress. Rising to her feet, her face the color of a fire engine, she yelled, "absolute twaddle, Feathersbee. There have been more great volcanic eruptions throughout geologic history than you have hair in your ears. In comparison, ice ages are extremely rare. Volcanic dust and gas may serve to lower the earth's temperature for a short period but not nearly long enough to cause an ice age. Clearly and absolutely, ice ages are due to the processes of plate tectonics. Ice ages are as rare as your logic, Feathersbee, because the earth not only has to be cold enough to cause large amounts of snow to fall and stay on the ground but, more importantly, it needs a large continental mass close to a polar region. The earth has to maneuver its jigsaw pieces to the right place before the ice giants come. Land—and the higher the land the better—must be in the right place. That is the key to an ice age."

"You always were a twit," yelled a small, young woman with long, greasy black hair. She wore a rather stylish black dress and white gloves that came up to her skinny elbows. Taking everyone in the audience by surprise, she leapt onto her seat and began speaking in a glacial voice. "You have it all teacups backwards. Ice ages, glaciers, cold; you constantly twaddle on about these past events like historians hiding in cobwebs of their own making. Really," and here she pointed toward the audience, "I'm ashamed of you all. It's the warming and melting of the ice you should be concerned with. That and the poisons we are putting into our air, the CO_2 and sulfur and the fluorocarbons eating up the ozone." Sweeping her arm around the room, cutting out all other ideas like bad weeds, she continued, "Blasted fools. Open your eyes and look. Ice age, smice age. The earth is getting hotter than a chili pepper, hotter than Madonna in a sauna, and just think of the consequences of that."

With the audience deathly quiet, she jumped down from her perch and marched to the front of the auditorium. Turning, she looked straight at a bearded gentleman in the front row and spat,

"You all know about fossil fuels. Such distinguished scientists"—and here she sneered—"should know that burning such fuels produces CO_2, an odorless, colorless, tasteless gas. The CO_2 gas traps radiant heat in the atmosphere and prevents it from escaping into the coldness of space; this is the so-called greenhouse effect. In a greenhouse, sunlight enters through the glass. It warms up the floor, the plants, the walls, the dog and the cat. These then radiate heat back through the glass into the great outdoors. Unfortunately, this heat is not in the form of visible light but is infrared radiation, and thus has a very difficult time escaping through the glass. Thus the greenhouse gets hotter and hotter.

"The earth could gradually heat up in the same way, and indeed, in the last 140 years, the level of atmospheric CO_2 has increased 25%.[4] If this rate of increase continues, then not far into the future we can kiss the polar ice caps goodbye. We could also say *arrivederci* to New York City, Boston, Tokyo, and Melbourne, to name but a few."[5]

"My dear young, ah, er, lady," replied the bearded gentleman in the front row. "Your point is well taken. But we know all this. Now," and here he paused for a second to look around the hall, a playful grin spreading across his face, "to mention a few bits of data you seem to have ignored. First, there is no clear relationship between global warming and increased CO_2 concentrations. CO_2 appears to go both up and down during warm periods.

"Second, detailed studies of growth rings from the 3,613-year-old alerce tree found in South America show no evidence that the climate is being warmed up. By measuring the widths of the tree rings seasonal rates of growth can be accurately determined. The rings show warm and cool periods over 3,500 years. They show absolutely no increase in growth rate since 1850, as you would expect if there were global warming.[6]

"Finally, the possible greenhouse effect you whine about may well be offset by the very air pollutants you decry. It is entirely possible that the soot, dust, and sulfur that is emitted into the air acts just like volcanic dust and gas: it reflects sunlight away from the planet. From this we get general cooling, and this counteracts the effects of increased CO_2. So quite possibly if you stop acid rain you will be starting your own great ice meltdown."

At this point, I am sad to report, the lady in the white gloves screamed some obscure obscenity (I think it was in Greek) and at the same time several other distinguished scientists stood up and started to shout angrily back and forth.

Thinking the time appropriate, I left the way I had come. As I closed the great doors all was in chaos; angry voices and shouting were mixed with breaking glass and pleas for calm and reason.

I wondered, as I left the building, if there was a final answer; would scientists ever agree to this cause or that effect? And what if they did? Would the earth stop having ice ages? Could we somehow prevent them? Getting carried away with my thoughts I asked myself, what if an ice age was the planet's way of saying to life, "you've gone and done it now, enough is enough—so I'm going to put your species on ice." It wasn't a nice thought, but as I stepped out into the noise and smell of the busy city street, I was sure I could hear the "glacier symphony" playing somewhere below me.

Prelude to Chapter 8

The sea; sometimes gentle and rhythmic as it meets the land, sometimes fierce and pounding. A gentle, flowing melody in the oboes as the sea slowly rises and falls, swell upon swell. Then the wind changes, the sea boils up, a crescendo in the full orchestra building to a "stormy" climax. Finally, a return to the gentle melodies of the oboes as the sea subsides and the waves again rub softly against the shore.

CHAPTER 8

Life on the Edge

Down to the sea
On rocky cliff or sandy beach,
That's where I'd rather be.

Many of us (me included) are drawn to the sea like moths to a bright light or, worse, to the all-consuming flame. What is it with the sea? What does it have we want or need or crave so badly? Possibly it's the sea's power, its force, its untamed rage.

The man, standing alone atop the scared, battle-weary cliff, faced the Atlantic storm. Shrouded in flying mist he was adrift in time, aimlessly tossed to and fro by 3.5 billion years of earth history, a history played out time and again in similar places.

Time and the shrieking wind. Furious it was with the stubborn sea and busy trying to drive masses of dark water landward. The sea resisted. For a heartbeat the water pulled back, then, swelling upward, it was swept forward on a last, mad run. Closing on the shore the boiling mass grew larger and darker, shaping itself into a towering menace topped by white, whipped foam. In a thunderous roar the wave hurled itself into the rocky cliff. The man could feel the force of the impact. The trembling cliff absorbed the blow. The wave broke, its spine shattering into an insane torrent of spray and pebbles. Rattling and banging, like a long, slow freight train, it came up the cliff.

The man never moved. Lifting his head he spread his arms wide and screamed his defiance at the angry sea. Laughing, the wind took his words and tossed them back to the boiling water. A second later the man was drenched by the rising spray.

Or possibly it's the pattern, the grace, the never ending comings and goings.

The great log must have been hurled onto the beach during last week's storm. It was an old tree, the rings told me that, but now it sat, an uncovered fossil, alone, and the perfect seat. Making myself comfortable on its scarred hide I stared into the vastness of the sea. The blue-green water was restless today. The swells, coming from who knew how many thousands of miles away, rose and fell in perfect rhythm, sure and steady, the pulse of the sea. Nearing shore each swell reared up, like a horse with a new rider on its back, to hang free for just a moment, before toppling forward in a sparkling shower of foam. A spent wave, replaced by another and another, on and on, in an endless parade. I sat there, hypnotized, unable to tear my eyes away; once again the sea had managed to hook my soul.

To many, standing on a wind-swept sea cliff, relaxing on or exploring a beach, or just staring out to sea, watching the comings and goings, is unsurpassed.

The sounds and sights of the sea lure us like the Sirens tried to lure Odysseus. So we come by the bunch to vacation, to visit, to build, to develop, to stay. On beaches, barrier islands, cliffs, spits, and anything else that will hold a structure and be within earshot of the rhythmic waves.

'Tis unfortunate, a cruel joke, that what sings to the deepest part of our souls is not our friend. Its song leads us to build on land that's as much of a geologic illusion as the magician David Copperfield's walk through the Great Wall of China. Oh the land looks solid enough; it's lovely and we much desire it. But to the sea and its master, the earth, the land, is history in the making. It is the land that is constantly being altered, submerged, or just plain washed away. Once again we are caught on the corkscrew of time. We think in days, years, and families; the sea thinks in decades, millennia, and storms. The sea and the dynamic earth; the coasts and us—sand and concrete. Here it is age, power, and the music of the earth that claims the victory.

> The shattered water made a misty din,
> Great waves looked over others coming in,
> And thought of doing something to the shore
> That water never did to land before.
>
> —Robert Frost

> Nothing under heaven is softer or more yielding than water, but when it attacks things hard and resistant there is not one of them that can prevail.
>
> —Lao-Tzu

A hard, resistant cliff, a beach of sand and gravel, a concrete-and-steel breakwater—these are but the dust of history.

Life is believed to have started in the seas. If it did, then millions of years ago our distant, water-dwelling ancestors turned their backs on the life-giving waters, and came to live on what to the sea is a four-letter word—Land.

Land that stands in the seas' rolling way; land that interferes with the freedom of its ebb and flow; land with whom the sea constantly wages an all-out, take-no-prisoners war. The land, when exposed to the sea, is immediately attacked by tides, waves, surf, wind, and storms. The land, it turns out, is a movable object sitting in the path of an unstoppable force.

In the face of this constant onslaught, coasts, beaches, cliffs, and islands are forever changing or just plain vanishing. Forty three percent of the United States' national and territorial shorelines are losing more sand, rock, and soil then they are receiving.

Yet people build and develop in this war zone, this zone of dynamic change. To see this change all you have to do is visit a beach after a storm, or the same beach weeks or months apart.

The tide is rising, the waves are small,
Was just the opposite when we were here last fall;
The beach is littered, full of shells and tiny bones,
Last time it was logs and polished stones;
And the sand has vanished from our picnic spot
Instead it's now but solid rock;
Wind and waves, roll and roar,
Come to change the beach for evermore.

The music plays on, oboes and orchestra in a dialogue of sea and wind, wild and crazy, forever at war with the shore. Those who live in this war zone really do live life on the edge.

SURF'S UP

Now the great winds shoreward blow,
Now the salt tides seaward flow;
Now the wild white horses play,
Chomp and chafe and toss in the spray.
— Matthew Arnold

Ever since we have gone to the sea in sailing ships and automobiles, to play and explore, we have watched waves, the majestic blue and green rollers that dance onto the shore to fascinate and delight, the angry gray and black storm waves that crash on the shore to scare and energize, and the joyful white, foaming breakers that climb up the legs of laughing children. We are born wave watchers, and this human habit goes deep to the core of our souls.

Water waves are not so much different from human waves created at football, hockey, and soccer games. Stand up, raise your arms, sit down, watch it go around. Stand up, raise your arms, sit down, watch it pass you by.

Waves are just that, the up-and-down motion that takes place on the surface of oceans, seas, lakes, ponds, rivers, and bathtubs. We call this up-and-down movement everything from ripples, rollers, breakers, and swells to whitecaps, seas, and surf. These are all waves, and waves are primarily caused by the wind, though rocks, boats, whales, and feet can form small, short-lived ones. Waves come in all sizes—toddler, petite, regular, big, tall, and extra large; there is no one size fits all.

To get some idea of how a wave forms, stand on the shore of a quiet beach on a day when the water is as smooth as a CD. If you can't find such a beach or such a day, a bathtub will do just fine. Locate a small rock and throw it into the water (if you're in a bathtub, dropping a bar of soap will work just as well); you have created instant waves. The rock caused a disturbance on the surface of the water, imparting energy to it, giving it life. This is similar to blowing on a bowl of hot chicken noodle soup to cool it off. The force of your breath across the surface of the soup causes it to ripple and move; the noodles are washed to the opposite side and waves are born. The chief wave making force of the ocean is the wind, whose breath constantly blows across the surface of the sea.

All waves, toddlers to extra large, are described in terms of their size and shape. They all have crests, or tops. Since they have tops they must have troughs, which are the valleys, or depressions, between the tops. The full height of an adult wave is the vertical distance between the top and valley bottom, and the wave length is the distance between successive wave tops passing a stationary point. Tops, valleys, heights, and lengths—the long and short of a wave.

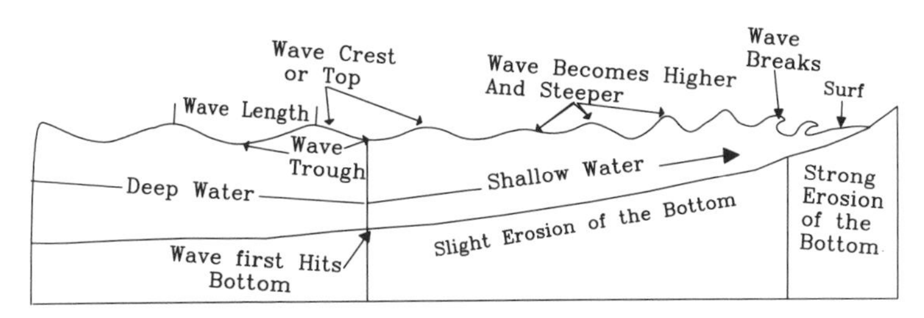

Parts of a wave, and changes that occur as waves move from deep into shallow water. (Adapted from B. J. Skinner and S. C. Porter, The Dynamic Earth, 1989)

In general, the height, length, and speed of a wave is dependent on the wind in three ways: (1) wind speed, (2) the length of time the wind has been blowing, and (3) the distance the wind has traveled without any land interference (also called the "fetch"). The greater these three factors are, the taller, longer, and faster the wave(s) created. In the open sea, normal winds, 2–15 miles per hour (3–24 kmph), form waves that range from 3–13 feet in height (0.9–14 m); storm winds can create waves from 10 to more than 100 feet in height (3.4–30 m). The largest wave ever recorded, which at sea means seen and white-knuckled through, struck the USS *Ramapo*, a navy tanker crossing the Pacific. During a severe storm the ship was overtaken by waves that, when measured against the ship's superstructure, were 112 feet high (34 m)![1]

Waves at sea are formed either by winds that blow fairly steadily, such as those of the westerly wind or the trade wind belts, or by local storms. Using the word *local* with ocean can be misleading because the oceans are so large that a "local" storm may cover an area of a few thousand square miles. Within a storm area winds are gusty and turbulent, so the waves that form are ragged and jumbled, a messy confusion of different heights and lengths. There is no regular pattern to storm waves, and sailors use the phrase "a real sea building" for these kind of waves.

Waves that manage to sneak out of the storm area become more uniform and regular—more civilized. These waves are lower in height and longer in length than storm waves, and that's just swell, or swells, as they are called. Swells can travel thousands of miles before they meet their nemesis—land. The swells that roll ashore onto the sands of St. Helena and Ascension, both located in the Atlantic Ocean far south of the equator, were born during storms in the North Atlantic. They travel over 3,720 miles (5989 km) to meet their fate, and we think salmon go a long way to die.

The normal fetch for a storm wave is about 500 miles (805 km), unless you happen to live in the roaring forties or ferocious fifties, a sailor's term for the latitudes of 40° and 50° S. Here the winds fetch storms for well over a thousand miles (>1,600 km), with nothing but water and ships to get in the way. In this manner great storms are born, brewed, and turned loose on the unsuspecting land.

If you happen to be a wave watcher, watching waves wash into shore, you can get some idea of their past history by their shape and pattern. If the waves are uniform in length and height then they are swells from a far-off storm. The so-called long-surfing waves that call in on the California coast in the summer were given life some 4,000 miles away (6,440 km) by winter storms in the Antarctic and South Pacific regions.

Waves that are different lengths and heights and break at various distances from shore, are the products of local storms superimposed on the swell of the sea.

How does a sea swell or run? It could be toddling along at a wimpy 2 or 3, or flowing at a blustery 8, or possibly rising with an extra large, ship-sinking 12. These numbers are not whimsical scores pulled out of a hat and delivered to skaters and gymnasts, but rather part of a numerical scale that sailors have used for more than 170 years to keep track of how the sea is behaving (or misbehaving). In 1806, Admiral Francis Beaufort, of Her Majesty's Royal Navy, developed a scale that allowed him to relate his observations on the state of the sea to measured surface wind speeds, wave shapes, and sizes. The Beaufort scale, which now ranges from 0 to 17, was adopted by the U.S. Navy in 1838. A 0 on the scale means the sea is like glass; a 4 indicates petite waves with whitecaps and a wind speed of 12 to 17 miles per hour (20–28 kmph). Getting into a size 8 is being fitted for what sailors call a "fresh gale": high waves, and foam and froth is tossed about with wild abandon; wind speeds range from 38 to 43 miles per hour (62–70 kmph). Moving over to the big and tall section we can try on a size 11: waves high enough to hide ships from view for long periods of time, and wind speeds are 66 to 70 miles per hour (107–113 kmph). Above 11 is the hurricane section, with waves bigger than huge, and wind speeds in excess of 73 miles per hour (118 kmph). No one wants to try on any of these sizes.

Not only does the wind drive water along the surface of the sea, it also imparts some of its force downward to stir and agitate.

Water being water, however, the water droplets bump and collide together causing what we can call "wet friction," which decreases the wind's energy as water depth increases. Measurements indicate that the wind can pester water down to a depth equal to about one-half of the wavelength at the surface; below this the water has a movement close to zero. The bad news, for those who like to swim with the fish, is that wavelengths up to 1,968 feet (600 m) have been measured in the Pacific Ocean. This means that the wind can toss and turn divers down to a depth of 984 feet (300 m). The good news is that most wavelengths are much shorter than this.

While waves remain where good waves should, out in deep water, they are unaffected by water depth because the base of the wave never comes close to touching the bottom. Moving into a new neighborhood, however, such as a posh coastal area, rapidly changes this situation. The up-and-coming wave is in for the shock of its young life. As the high-riding roller heads into shore, it eventually reaches a point where the water depth is about one-half its wave length. At this point the base of the wave collides with the bottom. The wave immediately slows down, like a motorist who sees a police car parked along the side of the road, and its shape rapidly changes. The wave height increases, while the distance between wave tops, the wavelength, decreases. This is similar to pushing a throw rug slowly into a wall. The carpet crumples into wavelike shapes, which get taller and closer together the more you push.

Waves in drag? You bet. Waves change their appearance as they advance toward shore because they travel at slower and slower speeds. This gives the waves behind the opportunity to catch up to the dragging waves ahead. Wavelengths thus get smaller and, just like pushing the carpet, the waves are pushed higher and higher. With the base of the wave dragging on the bottom, the high, steep top starts to travel faster. In fact, relative to the slowing base, it travels so fast it actually outruns the lower part of the wave. Suddenly, finding that there is nothing below for support, it topples forward and "breaks."

The turbulent water formed by this breaker advances up the shore and is referred to as "surf." Surf rushes up the beach or crashes into rocks until its energy has been drained away. Then, tired and worn out, it slowly slides back to sea in a broad sheet called "backwash." It is this backwash that creates the dangerous currents known as undertow.

When a wave falls apart, or breaks, it can go about it in three different ways: by spilling, plunging, or surging.

Spillers are like small children who have great difficulty walking and carrying a cup of apple juice at the same time. They slowly wobble and weave across the room, and the juice is placed in a cup without a lid (foolish parent), slops and splashes over the edge and down the child onto the carpet. As waves, spillers are slow wobblers, breakers that form where the bottom has a gradual slope. In this "gentle" environment the breaking wave slowly collapses, the top tipping over and down the front of the wave a little at a time. Because of this action, spillers are the waves that give surfers long, smooth rides.

Plungers are a lot like their human namesakes, reckless gamblers who rush ahead at all costs. Plunging waves form where the slope of the bottom is relatively steep. This being the case the bottom of the wave slows rapidly while the top keeps right on truckin'. Off it zooms into nothingness, and when its momentum can take it no further, down it comes in a rapid, crushing roar. Plungers give surfers short, but extremely exciting, rides.

Last, but not the slowest, are the surgers. These types of waves form where the bottom is so steep that the wave doesn't begin to break until it is right at the shoreline. There is a sudden increase in wave height and then, almost immediately, the wave topples forward to explode on the beach.

Waves, since they normally approach shore at some angle, won't touch bottom at the same spot along their entire length. Usually, the part of the wave nearest the shore will strike bottom first and slow down. This gives the seaward part of the wave, still traveling at normal speed, a chance to catch up. In this way the entire wave slowly bends or swings around toward shallow water,

Wave refraction at La Jolla Cove, California. Waves at the rocky point are at a distinct angle to the beach, while those coming ashore at the beach have been bent, or refracted, so that they almost parallel the shoreline. (Photo courtesy of T. Johnson)

forming a wave front that appears to be coming straight in. This process is referred to as wave refraction, the bending of a wave so that it aligns itself parallel to the shoreline.

Landmasses that project into the water, such as headlands and points, attract waves. The water around these features will be shallower than elsewhere, and so waves will naturally bend toward them. Consequently, wave energy will be concentrated on these features, leaving bays and coves in relative calm. In this way wave bending works to straighten out irregular coastlines.

SUDDEN IMPACT

A wave wants nothing more from life than to roll on, free and easy, across the ocean blue. When land gets in its way, there is only one thing for a good wave to do—give up the carefree life and go to work. Work for a wave means to move, crumble, erode, smash, gouge, submerge, and maim the land.

A wave does its best work during storms. A storm-driven wave can hurl thousands of tons of water and small stones against the land. At times this occurs with such force that the land really does tremble and shake, giving anyone standing on it the impression that an earthquake has tiptoed down the shore.

During a winter storm off the Atlantic a few years ago, a 1,350-ton portion of a steel-and-concrete breakwater that protected Wick, Scotland, was ripped from the structure and tossed ashore as easily as a chef tosses pizza dough. The good citizens of Wick would have none of that, so they replaced the broken portion with a new, improved 2,600-ton piece, a piece that would last forever. Five years later the sea took hold of this new piece and, ramming it right up the beach, gave it back to the good citizens of Wick.[2]

In 1952, during a fierce Atlantic storm, the bow half of a steamship was thrown up against a cliff so hard that it stuck like a dart in a dartboard. After the storm everyone came out to look at the bow, hanging some 148 feet (45 m) above the level of the sea.[3] Along with wave power comes rock power. Tillamook Light, off the Oregon coast, rises 131 feet (40 m) above the level of the sea. The light has had to have steel grates installed to protect the glass that shields it. The glass has been broken time and again by wave-thrown rocks higher, harder, and faster than a Randy Johnson fastball.

Wave power, rock power, and—look out—log power. Large driftwood logs, found on some beaches, can be set afloat by high tides. These then become battering rams, lethal weapons against the land when hurled shoreward by breaking storm waves. Even on days of calmer weather such logs signal potential danger. Logs are great to sit on and watch the sea, or to climb up and over and play on. Occasionally, a high breaker will come ashore and, with no warning, roll the log completely over. If someone happens to be on it at the time, he or she can be trapped underneath and face drowning or be crushed right then and there.

Rocks, logs, and water, the weapons of the sea, attack land in full force during storms, and on every coast, there is no shortage of these. It is no wonder that cracks, fractures, and breaks quickly appear in cliffs, sea walls, breakwaters, or anything else that is hard and resistant and set into, or onto, the land.

During a storm water is forced into every opening, no matter how small or tight. The forward thrust of the moving wave causes the air in the opening to be tightly compressed. When the wave has broken, its energy spent, the air rapidly expands, dislodging rock particles and larger rock fragments, which are then swept away by the next wave. In this manner the opening grows wider and wider and the rock, cliff, or seawall slowly crumbles and is carried away by the sea.

Waves, like their colder glacial cousins, also polish and abrade, though they do so in fast-forward speed. The pounding and grinding action of water, which contains more rock particles, pebbles, stones, and sand grains than New York City does rats, can do heavy duty damage to the shore. The smooth, shiny, round pebbles found along beaches the world over are a good reminder of just what one rock, grinding against another rock, can do. It is estimated that a 1-square-foot rock will become sand powder in about 100 years if kept in motion. Now what kind of person would bother to figure that out?

Along shorelines composed of unconsolidated material, such as sand, gravel, clay and/or glacial till, the rate of erosion can be faster than dancing the polka. Along parts of the coast of Great Britain, where the sea attacks deposits of sand and clay, the coast has retreated more than 3 miles (4.8 km) since the days of Hadrian and his wall. This rapid erosion has led to the disappearance of many old coastal villages, along with several well-known landmarks (at least to the Romans). A similar situation exists along the cliffs of Cape Cod, where glacial deposits are being removed at the rate of 3 feet per year (0.9 m). From the 700 miles of California coast, where erosion chews up 0.5 to 2.5 feet (15–75 cm) of land per year, eastward to the beaches of North Carolina, which have been worn back by as much 70 feet (20 m) during the past decade, the sea is reclaiming its lost empire. The historic light house at Cape Hatteras was built 5,000 feet (1,500 m) inland in 1879 to protect it from land erosion. Today it sits right at the shoreline! Tomorrow (geologically speaking, in hundred-year terms), it will be gone, along with many of the cottages, homes, and estates that, from California to Maine, share the same bird's-eye view of the singing sea.

The damage and destruction storm waves cause are mere child's play compared with the waves from hell—storm surges or storm tides. A 12 or better on the Beaufort scale, these storms pack extremely high water and high winds. They are centered around areas of very low atmospheric pressure, which causes the water under the center of the storm, the so-called eye, to rise up and form a huge, dome-shaped mass. Farther away, near the edges of the storm, the surface of the sea is actually depressed, like a fallen cake.

This great dome of water is carried across the surface under the center of the storm. As it nears the coast, water levels rapidly rise and, whipped by the high winds behind them, can cause extensive and disastrous flooding of shallow coastal areas. Smaller, if you can use that term here, storm surges may also form at the sea–land boundary. Here local high winds can drive hundreds of thousands of gallons of water into low-lying areas.

Once upon a time, on the North Sea coast in what is now The Netherlands, was a prosperous, fertile land known as North Freisenland, a large area covered with lakes and streams, farms and villages. Located on the River Hever was the city of Rungholt, the largest trading center in the northern region. The people living in this land were constantly at war with the North Sea. They had constructed many dikes and dams to protect their homes, villages, churches, and farms; until 1362 they had managed to keep the watery beast at bay. In that year, between January 15 and 17, giant waves, driven by high winds, drowned the land under a deluge of water. The moon was full, the tide was high, and this helped the flood waters rise higher than anyone could remember. Rungholt vanished in the storm, along with numerous villages, and 7,000 people drowned.

When the storm was over, and the sun crept out of hiding, all that remained of the land was one vast seascape dotted with a few desolate islands, now known as the Frisian Islands. This disaster of a storm was given the name *grote Manndranke* (great drowning) and is said to have caused the deaths of 100,000 people in the area between the Elbe and Repen rivers.

After the *grote Manndranke* only one large tract of land remained above the waters, a large island that came to be called the Strand. The Strand stubbornly stuck its chin above the sea until 1562, when the sea had enough and destroyed it. The Strand was flooded and eroded down to three tiny islands. In one night what had been a farming land of 50,000 hectares was reduced to three isolated bits with a combined acreage of 9,000 hectares.[4]

Lowlands facing the sea, to islands, to water—a natural evolution when the waters rise and the storms come.

In 1876, in the Bay of Bengal, a storm surge swept over the low-lying area and killed 100,000 people; in 1970, a similar surge struck the same area, claiming more than 300,000 lives.

In 1900 a storm surge created "the great flood of Galveston," Texas, which put the city under 13 feet of water and killed 2,000 people. More recently, Camile blew against Texas and Andrew drenched Florida. Imagine if Hurricane Andrew had tracked farther north into Miami or up through Tampa Bay. It didn't; hence, the Dolphins and Buccaneers play on, but so does the wind and the sea, and there may be a rematch.

As great land and coast blasters, storm surges may simply poop out, fizzle, or flop, which does not automatically mean that people and property are spared and safe. Such storms may bring torrential rains that cause rivers to rise well above their banks, setting the stage for biblical type-floods. Tropical storm Alberto was a fizzle as a land destroyer, but the rain he brought spelled disaster for the Flint River and Albany, Georgia, in the summer of 1994. Alberto created the worst flood in Georgia in 100 years, sending the Flint River more than 13 feet above flood stage. Thousands were evacuated, drinking water was contaminated, houses were buried up to their TV antennas, and 28 people lost their lives. Every year, somewhere near the sea, disaster strikes; as long as people live in a war zone, this will be the name of the game.

THERE IS A TIDE IN THE AFFAIRS OF THE SEA

The tide comes, stays never still;
It waits upon no prince's will.
—Motto on the lift lock, De Lemmer, Friesland, Netherlands

There is a tide in the affairs of people who live in coastal areas. That tide is the pulse of the living sea, the daily rise and fall of seawater around the edges of all the lands.

This daily change in sea level has been noticed and known ever since people first came to the sea; but the first person to relate it to something other than gods, magic, or great, mythical beasts was Pytheas of Massilia (Marseilles). Pytheas was a sailor and an explorer and so, around 330 B.C., he found himself off the coast of what today is Norway. There he observed the tide rise higher and fall farther then he had ever seen it do in his home waters of the Mediterranean. At the time of his observations the moon was full, a huge yellow orb hanging over the misty land. Pytheas, writing in his journal, wondered if such high tides were not caused by the attraction of such a full moon. Following up on Pytheas's act, both Aristotle, in Greece, and Pliny, in Rome, independently wrote that the seas responded to the double attraction of the moon and sun, but that the moon, though the smaller of the two, played a much larger role in the rise and fall of the tides.[5] To me this sounds an awful lot like the idea of gravitational attraction, without the exact words or the falling apple.

The Greeks knew, and the Romans knew; so why didn't anyone else know after the time of Pliny right up to 1687? Somehow the knowledge of the cause of the tides became obscure, forgotten, and lost; the good old Middle Ages strike again. Even the great minds of these times, Leonardo, Galileo, and Descartes, puzzled over and debated the cause of the rhythmic rise and fall of the seas. Then, in 1687, Isaac Newton had lunch in an apple orchard and the rest is history. Newton, while fiddling with his theory of gravity, rediscovered the cause of the tides. Writing on the "lunar theory of tides," Newton demonstrated that earthly tides are caused by the gravitational pull of the moon and, to a lesser extent, the sun upon the earth. The gravitational pull of the moon causes the earth to stretch in the moon's direction. Since our planet is composed mostly of solid things (like rocks, skyscrapers, and pizza), this stretching effect is ever so slight and largely ignored by most of the planet, with one big, nay giant, exception—the seas. Being large bodies of liquid and free flowing, they succumb to the moon's pull in a most dramatic fashion. They rise up to cover the land, then they fall away to leave the seafloor open and bare—"who has never seen the tides has never known the pulse of this good earth." Imagine if the moon had watery seas; what would their rise and fall be like under the influence of the earth's gravitational pull?

The tides come and go because of the changing position of the moon relative to the earth as the earth rotates through its 24-hour day. The effect of the sun is minimal, except when it is lined up directly behind the moon. Then the gravitational pull of the two bodies are added together to produce the biggest daily tidal range; at these times the tides are at their highest and also their very lowest levels. This happy event occurs twice a month, coming and going with the new and full moons. For ages these highest and lowest tides have been called "spring tides." They have nothing to do with spring (except some of them may roar in like a lion); instead, the rising water is said to "spring up, out of the sea."

The opposite of spring tides are neap tides. These occur when the sun is located at right angles to the moon. In this celestial position each body partially offsets the gravitational pull of the other, giving us tides that have the smallest range, or change, between high and low water. The difference between the springs and the neaps can be dramatic. Along the coast of Maine it is about 5 feet (1.5 m); at Pak-hoi, China, it is 16 feet (4.8 m); and on the wild, rocky coast of Brittany it is 13 feet (4 m).

Because of these large differences it has been said that "spring tides are beloved by all who live by or from the sea." The reason for this is that during a spring tide, low water exposes rocks, shoals, sandbars, tidal flats, and sea creatures that are otherwise not normally seen. For instance, there is a variety of seaweed that literally lives and dies by the rise and fall of the spring tide. This particular slimy, stringy, disgusting stuff needs to bask in the sun and breathe the good air twice a month to survive. If not for the spring tides it would no longer be with us. Low water during spring tides also brings out the clam diggers and crab chasers, this being their twice-a-month chance for delicious edibles.

The tide-producing force of the moon and sun is distributed evenly over the earth, but the differences in the sizes and shapes of ocean basins, as well as coastal areas, lead to large differences in tidal heights. Typically, tides range from 2 to 12 feet (0.6–3.6 m), but there can be as much as 46 feet (14 m) between high and low water; that's like the Jolly Green Giant standing beside a four-year-old child.

Tides generate a lot of excitement at the entrances to rivers and narrow bays. So much water in such small, shallow spaces so quickly is not good. When the tide rises, the first mass of water herded into a narrow opening is funneled through with a great rush, only to have the breaks applied as it strikes bottom. The water right behind quickly catches up and, in this way, a steep, moving wall is formed. The front of the wall is steep enough that water is constantly plunging down it, like breaking surf.

This watery wall, moving upriver or across a bay like the front line of football players moving upfield on a kickoff, is called a tidal bore. Most tidal bores are small, averaging less than 3 feet (0.9 m). However, there are a few giants around that are as high as 26 feet (8 m). Such great and famous rivers as the Trent in England, the Amazon in South America, the Seine in France, and the St. John between New Brunswick and Nova Scotia, Canada, have tidy little bores.

Tides to watch come and go, to expose flats so clams and crabs can be dug and to produce electrical power—all of these from the sea.

Tides have been a source of power for more than 900 years. Back in the Middle Ages people may not have had the foggiest notion about what caused them, but they sure knew how to use them. In coastal areas it was not uncommon to see waterwheels turned by the rise and fall of the tides. In the seventeenth and eighteenth centuries tidal mills along the European coast produced flour, lumber, and grain.

Tide power! Not the stain-removing kind, but the lighting up the town variety. Coastal areas with narrow rivers or bays and a large range between high and low tides, have the potential for the generation of electrical power from the sea.

To power up your TV or VCR, here is all you have to do. First, construct a dam across the river entrance, or on the land side of the bay or estuary. Second, construct a reservoir behind the dam to hold the water you collect during high tide. Third, have lunch while waiting for low tide. Fourth, at low tide release the high-tide water from the reservoir. The difference between the water levels allows the high tide water to flow outward with enough force to drive turbines and electric generators. Fifth, wait for the next high tide and repeat the entire process. To make certain your TV and VCR will get uninterrupted service, a tidal range of at least 23 feet (7 m) is necessary to provide a strong enough outflow. There are very few places in the world that have such aa tidal range, and even fewer with the proper narrow river or bay.

New! Improved! Extra Strong! It's the double-action tidal power plant, which operates in a similar manner to the above except, when the high-tide water is released, the reservoir gates are closed. The gates stay closed, and the water of the next high tide piles up against them on the seaward side of the dam. During the time of the rising tide the water level behind the gates remains lower. When the difference in water levels is deemed great enough, the gates are opened and the high-tide water rushes in to fill the reservoir and, on the way, powers up a generator or two.

Tidal power generation is environmentally friendly (unless you happen to be a clam or oyster living in the dammed estuary or river), clean, and economical. However, such power generation is restricted to a few isolated spots on the planet. There are only 255 potential sites, about 10 fewer than cities wanting their own football or basketball franchise. To supply the world's electrical needs, it turns out you would need 250,000 tidal power plants. 141 million windmills, a panel of solar mirrors that would cover all of North and South America, or 43 million elephants running in place on a treadmill would do the same thing. Aren't these cool tidbits for the next happy hour?

Currently, the world has three tidal power plants. One is located in France on the LaRance River Estuary. This plant generates 54,000 megawatts per year. Wow, now I'm impressed; but what does it really mean? The province of Nova Scotia has a plant in the tidal estuary of the Annapolis River that generates 30,000 to 40,000 megawatts of power per year. The last, if it is still operating, is a small plant at Kenlays Bay, Russia. In North America the Bay of Fundy has been investigated as a potential site on and off since 1930—apparently more off than on, for nothing has happened. If Quebec separates from Canada, possibly a free Nova Scotia and New Brunswick will think about it some more. In the United States, Passamaquoddy Bay in Maine, with a tidal range of 50 feet (15 m) has been on again, off again in this regard for over 50 years.

THE GHOST OF A COAST

The coast, where land meets sea, is defined as the area affected by waves and tides. Since waves are constant-motion machines, and tides are busy coming and going, coastal areas are always but a ghost of their former selves; they are forever being shaped and modified. The sea, however, does not run this remodeling business all on its own. The shape and nature of our present day coasts is helped along by other geologic events, such as changing sea level, the types of rocks exposed at the shoreline, and the kind of geologic activity taking place along the coastline.

Based on the above, coasts may be divided into two types: low coasts and high coasts.

Also called emergent coasts, high coasts, land rising up from the sea, form due to uplift of the land or a drop in sea level low coasts, land on the way down, are also called submergent coasts, form when sea level rises or land sinks.

Depending on the type of geologic activity, and the kinds of rocks exposed at the shore line, both high and low coasts may have nice beaches and tall cliffs. So, to tell a low, headed for the bottom of the sea coast, some of the features to look for are fjords (like those off the coast of Norway), estuaries (which are drowned valleys), barrier islands, spits, and sandbars that occur with wide, sandy beaches, like those found along the east coast of the United States. A high-coast environment is one of hills or mountains separated by deep valleys (Alaskan and British Columbian coasts), active volcanoes with the volcano building land seaward (Hawaii and Iceland), sand dunes deposited by the wind (western Sahara), and the rebound or uplift of land that has been freed from a continental ice sheet. In Hudson Bay, this glacial rebound is more than 0.4 inches per year (1 cm). In such places rebound has left old beach deposits more than 300 feet (91 m) above sea level.

The west coast of North America, where the Pacific and North American plates meet, where mountains rise and volcanoes stand on high, is an example of a high-coast environment. The east coast of the United States, with its estuaries, marshes, barrier islands and spits, where the land is low and at the mercy of the sea, is an example of a low-coast environment. What helps make and keep this a low coast is the constant rise in sea level caused by the slow melting (estimated at 1 foot per century) of the polar ice caps, melting due largely to increased amounts of atmospheric CO_2. No big deal, you say, just pass the jumbo dogs. OK, except that along coastal areas where the bottom slope is gradual or almost flat. In these place a small rise in the level of the sea translates into a large inland retreat of the shoreline. A one-foot (0.3 m) rise in sea level would be equal to 90 to 200 feet (27–61 m) of inland beach and coastal erosion. In other words, along the Atlantic Coastal Plain, in places like Pamlico Sound, North Carolina, the shoreline would shift about two miles (3 km) inland.[5]

UPWARDLY MOBILE BEACHES

Henry Beston said that the "three great elemental sounds in nature are the sound of rain, the sound of wind in a primeval wood, and the sound of outer ocean on a beach."

For most of us, our first encounter with the ocean is the beach. To a geologist, a beach is a gentle sloping shore covered by silt, sand, or gravel that is worked by waves and tides. To most everyone else a beach is a place of sand, sun, and fun; beaches equate to swimming, warmth, volleyball, tanning, and sandcastles. When we visit a beach seldom do we stop to think about where the beach actually came from or, more importantly, where the sand under our bare feet is going.

In the 1700s it was generally believed the sand found on beaches formed by crystallizing out of seawater, like sugar does from maple syrup. It was thought that sand was a type of salt, and that the sea produced it along its shores like it did other salts. Today we know that sand is ground-up rock, and that the waves of the sea are the chief grinders. Not only do waves grind rock to make sand, but they also sort the sand and rocks into piles that are about the same size. So we end up with fine sand beaches, coarse sand beaches, pebble beaches, and real lumpy rock beaches.

White sand beaches form largely from ground-up shells and other marine critters; yellow sand beaches, such as those of California, come from the grinding of granite; the grayish green beaches of the Pacific Northwest are ground-up lava; and the black sand beaches of Hawaii are ground-up pumice or finely pulverized dark minerals, such as magnetite and ilmenite, derived from the meeting of lava and water.

Beach drift—sand moving onto, along, and off a beach—is caused by waves. We already know waves arriving at the shore are not normal. They are bent or turned to give the appearance of coming right toward us. This is just an impression though, for all waves actually end up striking the beach at an angle, however slight it might be. The wave topples and breaks, water rushes up the beach at an angle, then, having spent the last of its energy, it is

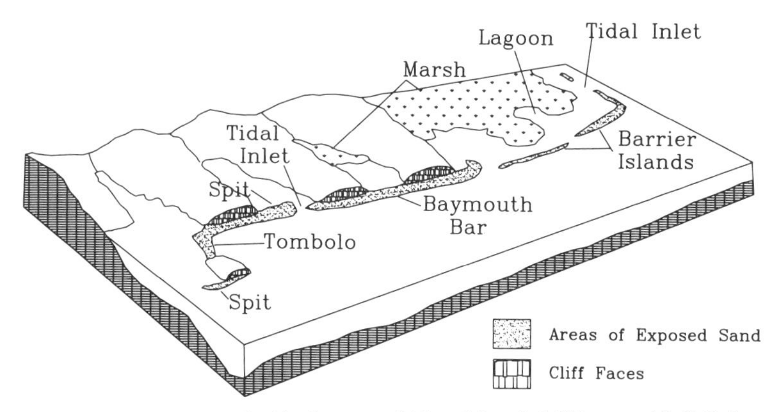

Geologic features associated with a low coast. (Adapted from B. J. Skinner and S. C. Potter, The Dynamic Earth, *1989)*

carried by gravity straight back down the slope of the beach. The effect of such an angled first strike, coupled with a straight backflow, is to move sand grains along the beach, much like a wide receiver in football runs a pass pattern—zig zagging from one end of the field to the other. This seemingly innocent movement can transport sand grains (and small pebbles) hundreds to thousands of feet in a day. The higher the initial angle of wave impact, the further down the beach the grain of sand will be moved.

Small pebbles that have been marked and tagged have been observed to zig zag along a beach more than one-half mile a day (0.8 km). Beaches are constantly on the move; the sand always seems to think things will be better on the other side of the point.

Waves striking the beach at an angle also form from currents in the turbulent water of the surf zone. These travel parallel to the shore line and, for this reason, are called long shore currents. Long shore currents move sand, gravel, and pebbles along the bottom surface and, when this material is added to that moved by beach drift, the total is mind-boggling. At Sandy Hook, New Jersey, the amount of sand transported along the shore over the past 50 years is estimated to have been 37.5 million tons. It is no wonder the

Sand spit off of New York City. The end of the spit curves landward in response to currents created by breaking surf. (Photo courtesy of R. W. Ojakangas)

poorly designed harbors fill up with sand faster than rats do on garbage.

Storm waves have the potential to move a ton more sand than regular waves. During a storm so much water is driven ashore so fast the beach sand becomes waterlogged. Once this occurs the water supplied by additional waves is shed off the beach like water from a duck's back. This means any sand the waves are carrying will be washed back to sea, never to be deposited on the beach. Storms thus stop the deposition of sand, but, being energetic and powerful, they can scoop wet sand up off the beach. Storm waves are all erosion and no deposition.

The interplay between long shore currents, beach drift, and moving sand leads to the formation of a variety of popular coastal features. Each of these features is much like the beach or beaches it was born from—just hanging loose, biding its time before moving on. Some of the more common of these hangers-on are the following:

Spits. These are elongated ridges of sand that project from land into the entrance of an adjacent bay. Most spits are continuations of beaches. They are built by long shore currents that dump the sand they are carrying at the very first place the water deepens, such as the mouth of a bay. Typically, the end of a spit will curve landward in response to the currents created by the breaking surf. Cape Cod is a wonderful example of a spit.

Baymouth bars. These are not places to drink on the beach, but are sandbars that extend completely across a bay, thus sealing it off from the sea. These form across bays where the long shore currents are weak, allowing a spit to grow and grow until it finds itself firmly fixed to the opposite shore.

Tombolos. Not strange musical instruments, tombolos are ridges of sand or gravel connecting an island to the mainland or to another island. They form the same way spits do. The word originates either from the Italian word *etymol*, which means sand dune, or from the Latin word *tumulus*, meaning mound.

Barrier islands. These are long, low, offshore ridges of sand that parallel a coastline. They form where the coastline is low and gently sloping, such as on the East Coast from New York City south to northern Florida. Barrier islands typically consist of a wide beach backed by sand dunes, and separated from the mainland by marshy lagoons. The lagoons represent areas of quiet water that allow small boats to avoid the open, choppier sea.

Exactly how barrier islands form is not really known. It is possible they are spits that have been cut off from shore by wave erosion or by the general rise in sea level that has occurred since the last ice age. They could also be sand dune ridges that were present along the coast during the last glacial period, when sea level was lower. As sea level increased, the low-lying areas behind the sand dunes became flooded, leaving the dunes separated from the mainland by low, marshy ground. Regardless of how they formed there sure are a lot of them, and because of our insatiable thirst for beachfront property, they have been extensively developed over the past 40 years. Two of the more famous barrier islands are Coney Island, New York, and Padre Island, Texas.

Unfortunately, these island paradises are shiftier than a politician up for reelection. Facing the open ocean, they take on the full anger and force of major storms. The islands are so low that during a storm, water several feet deep may wash completely over them.

When an Atlantic storm "comes a'callin," the island absorbs the force of the wind and waves by allowing itself to be freely shifted, molded, and moved, sort of like finger jello in the hands of a child. Sand is moved from the beach into the sea or onto the dunes; dunes can be eroded away and deposited as sand on the beach, or in the marsh or the sea. In general, since the sea attacks on the seaward side (that's a real revelation), the island retreats landward. Erosion on barrier islands averages 1.5 to 3.2 feet (0.5 to 2 m) per year, but can be as high as 65 feet (20 m), depending on the number of wild horses and how much grass they eat. That's right; believe it or not, according to some residents of these islands, expanding herds of wild horses are devouring the grasses that, for the moment, are helping to hold beaches and dunes in place. The problem has gotten so bad the National Park Service, with nothing better to do, is out with dart guns shooting mares full of contraceptives. Now don't you go laughing, this is serious business, especially when your front yard is vanishing faster than a Big Mac in the hands of a teenager. Taking proper care of those voracious, chomping, fornicating horses may add days or months even to the life of a barrier island.

These beautiful, sandy places, with the wide beaches and lovely sea views, with or without wild horses, are unstable. He or she "who builds his (or her) house on sands deserves a fool's cap."[6] Yet their lure has led to extensive building and development, all of which, along with the people who live there, is at risk. The people are not only at risk but also poorer by the pound (of sand), for they are constantly throwing dollars into structures and schemes to try to stop the sand from shifting. They build structures like breakwaters and groins, and they support beach nourishment programs. Over the years it has finally sunk in that beaches are dynamic places and that this kind of interference just robs Peter to pay Paul. Something that happens to, or is done, in one place affects the entire beach, leading to many new problems that are unwanted and difficult to correct.

Breakwaters are cement and/or steel structures built parallel to the shoreline to protect boats and beaches from the force of large waves. The idea is to create a zone of quiet water close to shore where boats can rest and the beach can sleep peacefully through the storm. As it turns out, not only does the beach behind the breakwater sleep peacefully, it also gains considerable weight. It grows larger, while the beaches on either side of the breakwater are more quickly washed into the sea. So here you are, on the marble terrace of your $2 million house, sipping martinis, watching your neighbor's beach blow by your breakwater.

Though Santa Monica, California, is not a barrier island, the story of its breakwater experience is instructive. The yachties (yuppies on boats) clamored for a breakwater to protect them from the cruel sea. And so it came to pass. The city built a breakwater parallel to the shore, figuring this would protect the boats, yet let sand drift right on through. The sand loved the quiet water behind the breakwater more than the boats did; it settled out faster than people in Congress accept free trips. As the sand settled to fill up the harbor, "downstream" beaches, considered to be "valuable" oceanfront property, were starved of sand and quickly gobbled up by the sea.[7] Lost beaches and constant dredging, all so yachties' sails wouldn't flutter in the wind.

Groins are barriers built at right angles to the beach for the purpose of trapping sand that is moving parallel to the shore in a long shore current. When the current strikes the groin it immediately deposits the sand it is carrying. The result is an irregular, but wider beach up current from the groin. Groins work so well that the current on the other side of the structure has no sand left in it, and so the rest of the beach starves. Sand is removed from these dying areas, and the beach vanishes into history. The solution (the Army Corps of Engineers must love this): build more groins. Groins to the left, groins to the right; yea, we shall fight the sea with our groins. There are more than 300 of these along the New Jersey coast, most built in response to previous groin construction.

The wisdom of building any of these structures has been called into question. Halting receding shorelines with protective structures benefits mostly those who have more money than brains, and it seriously degrades or destroys the natural beach, and the value it holds for the vast majority of us. These structures deflect the ocean's energy from a few places, only to refocus it on adjacent beaches and disrupt the natural flow of sand.

Beaches are nomads, have been nomads, and always will be nomads. Our attempts to shape them into our image of what we think they should be is a lost cause if ever there was one.

Probably the most acceptable method used to protect beaches is to feed a starving beach. Called beach nourishment, all it takes is the addition of tons of sand to the beach via La Dump Truck. By building beaches seaward the quality of the beach and storm protection are both improved. However, the diesel and gasoline transportation of sand from point A to point B is not only expensive but also only temporary. Virginia Beach dumps more than 30,000 truck loads of sand on the beach each year at an expense of more than $1.5 million.[8] Miami Beach funded 15 miles (24 km) of beach feeding at a cost of $64 million (how many refugees would that have housed and fed?). The original beach consisted largely of shell fragments. The sand added to the beach was made up of small quartz grains and fine clay particles. When wave erosion carried the clay out to sea, the sparkling blue water turned cloudy, and the nearby coral reefs were damaged (a coral reef needs clear, sunlit waters in which to grow). I also wonder what good the entire program would have done if Hurricane Andrew had decided to drift farther north.

Even with these types of programs, these communities, and many like them, are losing their sand grains to the sea. Not only is the daily wear and tear of the sea taking its toll, but the constant, steady, unstoppable rise of sea level is finishing the job. It is believed that sea level will rise another 16 inches (40 cm) by the year 2050. Small as this sounds, it will result in 10 to 10,000 times as much shoreline retreat and loss. Your beach or mine? Only time will tell.

A VIEW FROM ABOVE

The fierce Brittany coast is torn and ragged. For more than 600 miles (966 km) cliffs tower over the landscape, and caves, reefs,

Flat, beachlike "sea patios" cut into the rock by wave action. The cliff is slowly receding landward. (Photo courtesy of T. Johnson)

arches, and vaults echo with the music of the sea. Relentless, the sea has, for centuries, sent wild storms and mighty Atlantic breakers to split, crack, break, and remove the gray, red, and yellow rocks, crushing and grinding them to make beaches of sand.

It seems to matter not a tinker's damn whether the cliff is owned by Dover and made of soft, white chalk, owned by Lizard Point or Land's End and made of hard, dark mafic rock, or purchased by British Columbia and Alaska and composed of pink and gray granitic rock; the effects of wave erosion are the same and plain to see: a sea cliff is but a beach in the making. There are caves, rocky, narrow beaches, arches, sea stacks, and, of course, the high cliffs that give them birth.

Wave-cut cliffs result from the oversteepening of land due to erosion and removal of material from the base. Surf, wind, and waves pound into the base of a rocky shore or the bottom of a clay, gravel, or till bank, cutting, eroding, and carrying away. As the seasons turn and erosion progresses, rocks and soil that overhang the cut or notched base become unstable, and crumble and/or slide into the surf. In this way the sea constructs a cliff, and the land retreats. Some of the fallen cliff remains along the water's edge as part of a rocky beach, whereas the rest is swallowed by the sea.

As the cliff evolves and the land scurries for safety, it leaves behind a flat, benchlike surface called a wave-cut platform, which looks much like a loading dock or a patio deck. Cut across the bedrock by the surf, the sea patio slopes gently seaward and is continually being enlarged landward as the cliff recedes into the sunset. The shoreward part of sea patios may be exposed at low tide. If the coast is currently a high coast, then entire patios may be exposed above the level of the sea as the land rises or the sea falls.

Headlands, rocky ridges, points, anything that sticks out into the sea is ferociously attacked by waves and surf. These waves we can call the "wolves of the sea," for they attack the rocks of a headland or point by selectively wearing away the softer, more highly fractured, or weaker rocks, leaving the strong for a later day. As the weak and broken fall, sea caves are formed. When two sea caves, on opposite sides of a headland, unite, a sea arch is born. With years of wear and tear the arch will collapse, leaving a single, isolated remnant of the entire event sticking up out of the sea, like the an angry middle finger telling the waves where to go. Alas, it is for naught; even the stack will be sacked, and the whole history of this small war will have vanished.

Sand, clay, and glacial till are more easily disposed of than rocks. They undercut, crumble, and slide more easily; for this reason shorelines constructed of this stuff can virtually change overnight. During a storm in 1944, cliffs of sand and clay on Cape Cod retreated 16 feet (4.9 m); that's more than 50 times the normal annual retreat.

Various means, often as unsuccessful as they are expensive, have been tried to stop cliff erosion. Most of these are about as effective as me trying to stop Shaquille O'Neil from dunking a basketball. The most common method is similar to that used for beaches—build a barrier at the base of the cliff to break the force of the waves. This barrier can take the form of a concrete wall (breakwater or seawall), or a pile of large boulders called a riprap (which is exactly what the sea is going to do to the whole thing). If these barriers are constructed along only part of the cliff, the water sails around, behind, and under them as easily as Tinker Bell flies. The whole effort ultimately fails and, in the meantime, erosion continues full bore on either side.

EL NIÑO RIDES AGAIN

El Niño, "Christ child," is so named by South American fishermen because this phenomenon has a penchant for showing up during the season that Handel's *Messiah* is usually performed. El Niño, much like Mr. Scrooge, brings no joy or presents, but instead spells unpleasantness for all around it—bah humbug. El Niño is a warm current of water that effectively replaces or prevents colder, nutrient-rich waters from rising to the surface. The result is starvation and massive carnage among fish, plants, and birds.

The waters of the world's ocean basins are zoned vertically with respect to temperature. Near-surface waters (down to 300–650 feet [100–200 m]) are stirred, not shaken, by winds and waves, which means they are well mixed, oxygenated, and relatively warm. Within this layer, professionally called the photic layer because it is light penetrated, lives virtually all of a sea's plants, and much of its other swimming and crawling and floating forms of life. Below the living layer lurks a colder zone that extends down to a depth of about 3,280 feet (1,000 m). This is a zone of gradually decreasing temperature called the "did you bring your long johns" zone, or the thermocline. Temperatures at the bottom of this layer are about the same as the average annual temperature in Duluth, Minnesota—40° F (5° C). The deepest, darkest layer is also the coldest. Below 3,280 feet (1,000 m), water temperatures are 32° to 35° F (0–1.6° C), and the water is slow and heavy. Even though temperatures are at the freezing point, the water remains all liquid because of the salt content and the relatively high pressure. This frigid layer forms in polar regions and, being cold and heavy, it sinks straight to the bottom where it begins to migrate outward. It moves south, toward the warmer climates of the temperate and equatorial areas, moving so slowly that it may take centuries to travel from pole to equator.

Three zones, all neatly stacked one upon the other like pancakes, until the trade winds blow. Steady winds of the trade zone blow seaward, sending the warm, surface waters out on a long voyage aboard the love boat. As the warm waters slip away, denser, colder water rises from the "long-john layer" to take its place. This water is beloved by all who eat and sleep in the sea. Not only is it cool and refreshing, but it is also packed full of the good things necessary for life—nitrogen and phosphorous.

When this nutrient-laden water rises into the surface zone, it begins to warm, and is able to support abundant plant growth which, in turn, attracts many marine life-forms. Some of the world's greatest fishing grounds occur in these zones of upwelling.

So what happens if the trade winds blow the wrong way? What happens if the cold waters don't show up for their expected engagement? The warm, surface water stays, the colder, full-of-goodies water remains put, and El Niño decides to visit. This disastrous event appears to occur about every 5 to 10 years, but exactly what causes it, and why some El Niños hang around forever and others just pop in and out, is not really known.

The most popular theory going the rounds today is that El Niño is caused by what is known as the "Southern Oscillation." So all you northerners planning to go south for the winter please stay home. Actually, this event is a reversal in the way the trade winds blow, which seems to be somehow connected to the changes in atmospheric pressure in Australia and South America. Really, no joke. This is not the plot of a James Bond movie but an actual fact first observed in 1924. When there is high pressure in South America, like when Brazil was playing in the final of the World Cup in 1994, there is corresponding low pressure in Australia, and vice versa. These pressure reversals are related to changes in the pattern of ocean currents of the equatorial Pacific. When the trade winds blow true (off of South America), the warm, surface waters are carried out into the western Pacific. When the winds are reversed the warm water either stays put, or flows back into the eastern Pacific to depress the cold water and bring El Niño to town.

Significant El Niño conditions occurred in 1957–58, 1965, 1972–73, 1982–83, and 1991–92. The El Niño of 1982–83 increased surface-water temperatures off the coast of Peru enough to destroy the Peruvian anchovy industry. The anchovy catch dropped from 12 million tons in 1970 to less than ½ million tons in 1983. Dead anchovies means not only that pizza with the works goes way up in price, but also that other fish are killed off in large numbers and, because of this, thousands of birds starve to death. It has been estimated that more than 22 million birds perished from the El Niño of 1957. El Niño is also believed to affect more than life in the sea. It is also blamed for large, unusual changes in weather conditions. The El Niño of 1982–83 ushered in severe droughts in Australia and Brazil, terrible brush fires in Australia, severe storms on the West Coast of North America, and an unusually warm winter in Europe that had all the skiers hopping mad.

WAR AT SEA

We started this chapter by mentioning that the sea wages a constant war on the land and on all those who dwell in coastal areas. In the last 50 years coastal dwellers have brought the war to the sea, to the benefit of no one, and to the potential harm of everyone.

In the United States, 50% of the population lives within 50 miles of a major coast (this includes the Great Lakes). These millions of people, with their power and waste disposal requirements and heavy industry, have created a tremendous problem for the fringing oceans and lakes. At the turn of the twentieth century and on until the end of the Korean War, there was a naive belief that the oceans were indeed magic waters, that what got chucked into them would vanish forever. All the garbage and wastes would be dispersed, purified, and cleansed; the ocean waters had an infinite capacity to absorb and remove the byproducts of human occupation. The dream ended when the sea began returning these "disposables" to the land, and when the ocean waters became so polluted that fishing and the collecting of shellfish had to be curtailed and then stopped. It was too many people, too much use, and too much waste dumped into the waters at too high a rate. In early September of 1994, volunteers scoured more than 4,500 miles (7,245 km) of U.S. shoreline, collecting some 7.3 million pieces of trash in just 3 hours. The garbage collected ranged from pop cans and condoms to syringes. It included 344,502 pieces of glass, 331,963 bottles, 210,553 cans, 30,326 light bulbs, 55,430 trash bags, 1 human skeleton, and a 5-pound bag of cocaine labeled "radioactive substance."

The seas and the life they nourish are in trouble because of us. This plight of the sea has been documented in many fine books, in articles, and in television documentaries. This being so, we will turn our attention to two of the rather more disgusting examples.

At the mouth of the Hudson River, between clean, tranquil New York and pastoral New Jersey, sits the New York Bight[9] ("blight" would be a much better word for it). Garbage, the refuse of the streets, chemicals, dredge slop, and a body or two have been dumped into this "paradise lost" since 1890. Some of the garbage found its way back to the beaches, so in 1934, laws were passed prohibiting the dumping of so-called floatables in the bight. After the laws were passed you could hear the laughter of organized crime clear across town. Even with "floatables" denied access to the bight, toxic wastes, chemicals, and raw sewage were considered safe enough to deposit. The philosophy seems to have been, if it doesn't glow it won't bother or harm you.

It is estimated that from 1890 to 1972 the volume of waste dumped in the bight was 1.6 million cubic metric tons—another of those easy-to-visualize numbers. Is this a football field full, a stadium full, or more? It turns out it is a lot more. How about enough garbage to cover all of Manhattan Island to a height of a six-story building (66 feet, or 20 m). What a wonderful way to go. A lot of the stuff dumped into the bight was toxic and oxygen using, so the only forms of life left in these waters are mutated alligators (evolved from beloved pets flushed down the toilet by caring owners) and amphibious sewer rats.

Onward to the glowing poison of Minamata Bay, Japan. The people living there don't need a Red Lobster restaurant; they harvest their own shellfish, and they love to eat them fresh and in great abundance. Since they collected them during the day I guess they didn't notice their shellfish glowed in the dark. Between 1953 and 1960, a local industrial plant released high levels of mercury (the toxic stuff Minnesota banned from glow-in-the-dark tennis shoes) directly into the coastal waters. The shellfish sucked it up and then passed it on to the people who ate them, leading to mercury poisoning and numerous deaths. The physical and mental degenerative effects were especially severe on children whose mothers had consumed large amounts of the shellfish during pregnancy. The debilitating effect of this mercury poisoning has come be known as Minamata disease.[10] When local governments warn of eating too many fish because of the danger of mercury poisoning, believe them.

Finally, there is a red tide rising. I think Nikita Khrushchev said this about the influence of communism on the world stage. Today the rising red tide is no longer directed by the "old" USSR but by the sea. Red tides occur when either natural or human factors cause a rapid increase in the production of dinoflagellates. These are not dinosaurs whipping themselves into a killing frenzy, but tiny (<0.1 inches [0.01–0.25 mm]), single-celled organisms that just love to snack on water rich in nitrogen and phosphorous. Overall, dinoflagellates are a lot like wild mushrooms; some are poisonous, some are extremely poisonous, and some are perfectly safe. The poisonous types can cause such delightful illnesses as paralytic shellfish poisoning, or neurotoxic shellfish poisoning in people who dine on mussels, clams, and oysters. The toxins produced by dinoflagellates become concentrated in the tissue of shellfish as they devour these sumptuous tidbits by the thousands. The toxins do no harm to the shellfish, but when eaten by humans, they react with stomach acid to make a Molotov cocktail. Cooking your clam florentines well won't help either, for the toxins are unaffected by heat. Some of the poison produced by these tiny creatures is 50 times more powerful than strychnine. Arsenic and steamed clams, anyone?

Red tides have been known since biblical times but seem to be getting more and more common. Sewage and runoff from farms, lawns, and laundromats contain lots of nitrogen and phosphorous. When this is added to the natural quantities found in coastal waters, a mighty population explosion can be set off and the sea surface turns into dinoflagellate city. Quick as a cell can divide the dinoflagellates spread out, like a royal red carpet, across the water to be ingested by hungry shellfish. A red tide in the fall of 1986 spread along more than 300 miles (483 km) of the Texas coast; three quarters of the coast south of Galveston was closed to the shellfish business. During this red tide, business at coastal oyster and seafood processing plants dropped as much as 98%.[11] It is wise to also remember that even when the water returns to its normal blue-green color, the shellfish can retain the toxin in their tissues for long periods of time.

In the end, it comes around to these questions. To swim or not to swim? To fish or not to fish? To harvest or not to harvest? The answer from the sea is "what, me worry?" For us it may well come down to "to be or not to be?" The oboes, accompanied by the full orchestra, will play on, regardless of the amount of garbage, oil, and toxins we dump into the blue-green waters. However, like a very bad Broadway musical, they may well play to an empty house.

Prelude to Chapter 9

The desert: heat and the constantly blowing wind, a land of shifting sand and ageless rock. The music of such a place begins with a wordless chorus, the constant hum of the desert wind. Introduced into this background are high, glassy harmonics in the violins. As the violins play, there comes a swelling crescendo of the chorus and a sandstorm is born.

CHAPTER 9

Summer's Fire

Dust on the waterless plain blows over his tracks,
The sun glares down on the stones
And the stones glare back.

—*Stewart Douglas*[1]

It wasn't supposed to be like this. Not his homecoming. Four years of hard labor in the oil field, passing much of the time between shifts thinking of home and her. Thinking of his return, the celebration, the joy, the hard-earned money for family and marriage. All gone to sand in the heat and wind.

"Damn it. Damn the sand," and he kicked a small pile into the air. Instantly the wind reached out and snatched it away, just like it had his dreams.

"Damn it all to hell."

Hell, that's what his home, the Sahel, had become. Yet, standing here on the edge of the plateau, he remembered differently. Cows, hundreds of them, grazing on clumps of grass or on the low, broad-leafed Had bushes. The slope to the plain had been covered with shrub wood and long tufts of yellow grass and, after a particularly heavy rain, with pink and blue flowers.

The village began where the slope ended; from there it sprawled outward, a haphazard collection of huts, lean-tos, pens, and small gardens. The village existed because of the water, enough for everyone; cool and clear with just the hint of sulfur. He could still see where the wells had been, but they were more than wells, they were places where people gathered to smoke and talk, where children played and everyone laughed.

"Damn the sand." The plateau was horrible, grotesque, littered with dark green rocks that stood above the moving sand like rotting toadstools. The sand was pale red in color, as if it had soaked up the blood of his people, and that blood had spread everywhere; it was in ripples of sand, dunes and mounds of sand, into a sea of sand that rose and fell in the hot, freakish air like the heaving sides of a great, running beast.

Looking off to the south, he could see a moving column of dust devils. They twirled and twisted round and round, like merry-go-rounds, spinning in unison, reminding him of tall girls dressed in white and black, bobbing and turning to the beat of the hand drums.

"Stop it!" he told himself. This wasn't helping. He knew what must be done and he best be getting on with it. The floor of the plain, between the dunes, would be hard packed where the wind had swept it clear, and he had plenty of gas. He would follow the sand south; south to where it ended and his family began.

It had been four years of summer fire. Grassland to dust to desert. Drought, sand, and the hot wind had come, creeping and crawling like sun shadows, covering the land, its villages and people. Long ago the desert came because the earth spirit called it; it came with climatic change and creeping continents. In more recent times it has come by the invitation of humans; it has come because of the axe and shovel, the tractor and plow.

HOW DRY I AM

Deserts are arid lands, places like the Sahara, the Kalahari, the Great Victoria, and, closer to home, Death Valley. These are regions of heat, sand, dryness, bleached bones, and circling vultures. Areas without welcome mats that, next to the polar regions, are the least friendly lands on this earth.

A desert is defined as a land area where the annual precipitation is less than 10 inches (250 mm), and/or where the rate of evaporation exceeds the rate of precipitation.[2] These are places of so little vegetation they are incapable of supporting human populations (though certain spiders and lizards do quite well, thank you all the same). Deserts compose 25% of the world's land area outside of polar regions, with the largest of desert lands being the Sahara, which covers some 3.5 million square miles (9 million km^2).

Based on what they look like, their physical appearance, deserts can be divided into two main types: Olive Oyls and Mae Wests.

Everyone knows what Popeye's girlfriend looks like; she's skinny and long-legged. Olive Oyl deserts are as flat as pizza crusts and horribly long—mile after mile of flatness. Driving across one would be like driving across Lake Superior when it was completely frozen over, or Iowa, 50 times. The Sahara, Kalahari, and the great Australian deserts are Olive Oyl deserts to the nth degree. These deserts form in geologically stable areas, where the only earth music, over millions and millions of years, has been supplied by the wind and the shifting sand. Such deserts are dominated by erosional surfaces carved into, and out of, old, weather-beaten rocks.

"Next time you're around, honey, stop in and see what kind of shape I'm in," says a Mae West desert, with a shape that is the very opposite of flat and straight. Mae West deserts are dominated by steep-sided mountains separated by narrow basins. The basins are said to be enclosed, or interior, basins, for the water that manages to find its way into them never, ever comes out.

Mae West deserts, also called basin and ranges, occur in areas where the earth's music plays loud and strong, a symphony of volcanoes, earthquakes, flash floods, and the ever-present wind. Much of this music is meant to accompany the uplift of land caused by the meeting of two jigsaw pieces. Death Valley, the Mohave, and the Sonora deserts are well-shaped examples of Mae Wests.

The main agents that shape and manufacture Olive Oyl and Mae West deserts are the very same as those employed by the sea—wind, water, and sand. These three work together to provide the unique features of the desert landscape that impress, inspire, and bring thousands of visitors: features like sand dunes, barren rock, dry stream beds, green oases, sand seas, angular hills, steep canyons, and sparse, but often exotic, vegetation.

Both Olive Oyl and Mae West deserts are devoid of rain. The conditions that cause this perpetual thirst allow these two desert types to form in widely separated and very different places:

1. Having a prunelike face, that is, being continuously shriveled up and dried out, like my Aunt Blanche. Prune-faced deserts are areas of semipermanent high pressure, which greatly reduces, or actually stops, cloud formation. These areas have high temperatures and, therefore, high rates of evaporation, making it virtually impossible for any kind of self-respecting vegetation to make a decent living.

The Sahara, Kalahari (Arabic for "always dry"), and the great Australian deserts are examples of such dry places. In the prune-faced desert world of the southwest United States, the two driest spots are Death Valley and the Baja California area, both receiving less than 2 inches (5 cm) of rain a year. The longest "desert" drought in the United States was recorded at Baghdad, Arizona in the Mohave Desert, where the only measurable rain from 1917–1920 was $\frac{1}{100}$th of an inch. It is no wonder such places make your average prune look succulent and juicy.

2. Being a lonely heart, a shy wallflower located deep in the interior of a large landmass. Lonely heart deserts are far removed from the earth's principal sources of moisture: oceans, seas, and great lakes. The farther an air mass has to travel over dry land to visit a desert, the greater the chance of its losing the moisture it carries. Lonely heart deserts are visited largely by dry air. The Gobi Desert in central Asia is a good example of this desert type.

3. Being a shadow land, a hidden place that just won't get out on its own. These deserts are forever wanting to remain in the background, in the shadows. In this case the shadows happen to

be mountain ranges that stand in the way of moisture-soaked winds blowing in from the ocean. Wet air meets mountain and tries to climb up and over its top. As the air rises it turns cold and heavy, and this forces it to rain or snow over everything. Reaching the top, the air finds itself worn out, dried up; it has no moisture left for the land sitting in the mountain's shadow.

4. Being a seaside resort, which means these deserts may be the best of the worst. These are honest to goodness hot, dry, duney places right on the beach. They occur on the edges of continents where cold, upwelling seawater cools warm air blowing onto the shore. This cooling effect decreases the ability of the air to hold moisture, so it comes ashore about as dry as a W. C. Fields martini—the ones he made by pouring 2½ ounces of gin over ice, and then slowly waving the top of the Vermouth bottle over the glass.

The coastal deserts of Peru, Chile, and southwest Africa all sit at the edge of the sea but, like the ancient mariner, they are some of the driest places on earth. In the Atacama Desert of northern Chile, it can be as long as a decade between measurable precipitation.

Within both Olive Oyl and Mae West deserts, the amount of solar heat that radiates off the sand and rocks is so great that there is very little of it left by the time the sun goes down. This makes a desert a very cool place to visit after dark. The difference between late afternoon and evening temperatures can be as much as 90° F (32° C). No wonder plants, and many animals, need to be specially adapted to live here.

PAINTED DESERTS

When I think of a desert I think of sand. To many people the two are inseparable. Desert and sand go together like pizza and cheese, but they shouldn't, you know. They really should go together like pizza and anchovies, for sand makes up only a small percentage of most deserts. In the Sahara sand covers about 15% of the surface (of course, this is an area about equal to all of Alaska), and in the Arabian Desert, the world's sandiest, it covers 35%. The most common desert surfaces are barren, wind-swept rock or hard-packed, stone-covered ground.

Many of the barren rocks, as well as the stones on the ground, look as if someone came along with a paint brush and varnished them a shiny black or deep red color. This paint job is due to a thin coat of manganese and/or iron oxide that forms on the rocks after long exposure to the wind. So-called desert varnish is "painted" on the rocks by desert dust, which over the years and centuries settles onto the rocks. Small amounts of iron and manganese in the dust are washed out by summer rainstorms, to be immediately reprecipitated as black manganese or reddish iron oxide. This whole process may be helped along by microorganisms that live on the rocks. Desert varnish covers 75% of the barren rock in desert regions of the southwestern United States.[3]

THE WAYWARD WIND

The desert wind may have been born as a sea breeze, given life rattling around a mountain pass, or come of age in the highlands. Wherever it was born, it is relatively cool air drawn into the desert to replace stale air that has been thoroughly roasted and is now rising, like a hot-air balloon, upward and outward.

This drawing in, heating up, and ballooning out goes on constantly, which means the desert winds, in both Olive Oyl and Mae West deserts, blow from sunup to sundown, and often right on through the night. Constant wind does a number of evil deeds, such as drying out vegetation and soil, covering large areas with fine sand, and stripping millions of tons of sand and dust from desert areas each year.

Wind, like moving water, picks up loose things and carries them to other places. Compared to water, wind is a real lightweight, capable of carrying only fine materials, like sand, dust, and camel dung. On the other hand, wind is not bound by shores or channels, so it can spread the material it carries over large areas.

Sand grains, carried by the wind, are too heavy to be continuously suspended off the ground, so they move by hopping or bouncing along like ping-pong balls. This ping-ponging process, in your official scientific guidebook, is called saltation (from the Latin word *saltare*, which means "to jump").

Sand begins its journey by being rolled or pushed over the ground. When the rolling grain bumps into something, like another sand grain, a rock, or a camel, the grain can jump into the air. Once aloft it will be carried along, free of charge, until gravity pulls it back to earth. Hitting the ground, the wounded grain can stagger back up, like a fighter down for a 9 count, or scatter a whole bunch of other grains, sending many of them skyward. This starts a chain reaction, and within a very short time, an immense amount of sand can be picked up and moved.

"Sandstorm," he yelled into the camera. With a wave of his arm 100 tympani burst forth in sound, and the big cloud rolled forward. What you see in the movies ain't what it seems. Sand grains, when picked up and moved by the wind, rarely rise more than 3 feet (0.9 m) above the ground, regardless of wind speed. This is shown by telephone poles, fence posts, and the hairless legs of camels that have been sandblasted up to this height, but no higher. Sandstorms are mostly dust storms containing particles of silt and clay that can be swept high into the air. Being light, and having flat surfaces, they can stay aloft for a long period of time. Dust from the Sahara has been carried all the way to the West Indies, and dust from the 1930s "dust bowl" was carried over New England and out into the Atlantic.

Wind coaxes sand into the air by rolling and impacting it; dust, however, is too light for these shenanigans, and blowing on it to get it aloft works about as well as blowing on a kite. Sit by a dry, dirt road on a windy day and watch what the breeze does to the loose dirt—nothing; no dust rises from the road unless a rabbit hops across it. Let a logging truck roll by, however, and not only are you buried in dust, but a billowing cloud also rises upward to hang in the air.

To get dust into the wind it has to be disturbed, troubled on the surface. Something, like a camel, a rabbit, a truck, or, in the desert, a sand grain, has to come along and kick it up.

What all this means is that in deserts there are sandstorms below 3 feet, and dust storms the rest of the way up.

DEFLATION AND THE WILLIE-WILLIES

Wind, compared to waves, running water, and moving ice, doesn't have much fizz or pop as an agent of erosion. Even in deserts, where it is a force to be reckoned with, intermittent rainfall erodes faster and deeper. This is not to say the wind has no voice. On the contrary, not only does it erode and deposit material; but it also plays a musical instrument. The wind plays an aeolian harp, a wooden box with eight to fifteen strings, each of a different thickness, stretched along, and raised by low bridges above, the top. The instrument produces soft "exotic" sounds when the wind blows on the strings, causing them to vibrate.

The processes of wind erosion and deposition are called eolian. Add an *a* in front of the *e* and you have the word *aeolian*, after the Greek god of winds, Aeolus, for whom the aeolian harp was named.

Wind taketh in two ways: by a process called deflation and by abrasion, the same art form practiced by glaciers and waves.

Deflation is the lifting up and blowing away of loose things, like sand grains and silt and clay particles. The effects of this process are difficult to see on a day-to-day basis, because the entire surface is being lowered a little at a time. Overall the amount of erosion can be immense. During the Great Plains' dust storms of the 1930s, large areas were deflated by more than 3 feet (0.9 m) in just a few years.

The most visible result of deflation is shallow depressions called "blowouts." There are thousands of these from Texas north, across the Great Plains and on into Canada. Blowouts vary from the size of a circus tent to that of Texas stadium, with the rare one

Beginning stages in the formation of desert pavement. (Photo courtesy of C. L. Matsch)

approaching the size of the state of California; their depth varies from 4 to more than 160 feet (1.2–49 m). During wet years blowouts may be covered by grass, and can even contain shallow lakes. During dry years the water evaporates, grass dies, and the wind scoops up the soil and blows it away. Eventually these basins are deepened enough to reach the water table. Once this happens erosion comes to a standstill. At the water table the ground is moist and the soil particles will stick together, preventing the wind from moving them.

Deflation also paves the desert. Desert pavement, or as they call it in Australia, the gibber, is due to the blowing away of sand and silt, leaving larger stones behind. As the years roll by the desert floor ends up with a continuous cover of cobbles, blocks, and pebbles that may end up fitting together like the cobblestones that were deliberately placed on a San Francisco street. Once the desert floor is paved in this way, wind erosion comes to a halt.

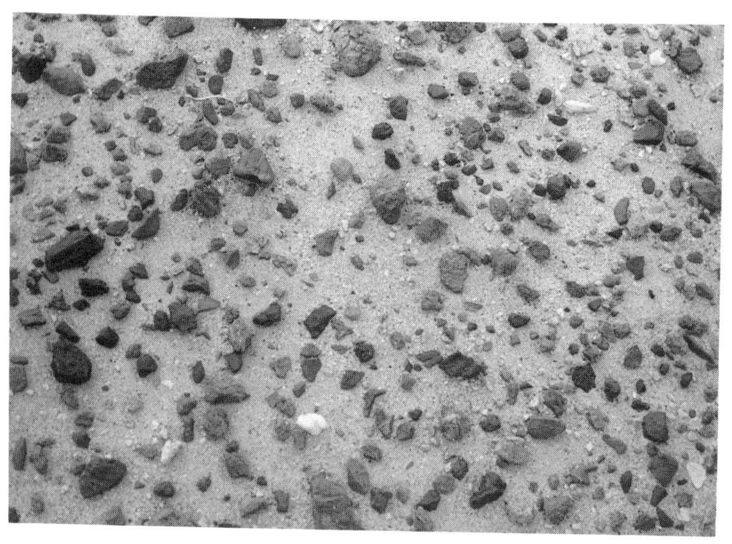

Desert pavement in intermediate stage. As the years flow by, the wind slowly blows away sand and silt-size material, leaving the larger stones behind.

End stage of desert pavement formation; the desert floor now has a continuous rock cover. Dark coloration of some of the rocks is the result of desert varnishing. (Photos courtesy of C. L. Matsch)

Ventifact with apple. Ventifacts are formed by wind abrasion. The rock in the picture has smooth, sharp edges that resemble arrowheads. The apple is for scale as well as for eating in the desert heat. (Photo courtesy of C. L. Matsch)

Desert pavement was a source of material for the stone tools of early humans, and was the origin of graphic art. Pavement pictographs, made by removing selected rocks from the pavement, are everlasting art forms that can be observed today in several different places. In the Panamint Valley of California, for instance, stone alignments of the desert pavement are believed to be more than 1,000 years old. In Nasca, Peru, pictographs over 4 miles (7 km) long remain as clear today as when they were first done, more than 2,000 years ago.[4]

The wind's other erosional media is sandblasting, or abrasion. Sand and silt, carried by the wind, impact into rocks, shrubs, and goats, slowly wearing them away. Abrasion during dust storms can strip paint, pit glass, erode rocks, and shear sheep. Wind abrasion also leads to the formation of ventifacts, rocks that have been polished and pitted by wind-blown sand. The rocks slowly become smooth and shiny, with sharp, arrowheadlike, edges.

Where strong winds attack unconsolidated material or layered rock, with one layer much tougher than the next, a yardang or two may form. Yardangs are elongate, streamlined ridges, much like glacial drumlins, that can be up to 328 feet (100 m) high and a few tens of miles long.

Desert pavement, yardangs, and ping-ponging sand grains are enough to give one the willie-willies. Actually, I hope not, for in the United States, Australian willie-willies are called dust devils, whirlwinds of dust and sand that form on relatively calm days when the surface of the desert has been intensely heated. Dust devils are formed from the rapid, and erratic, uprush of very hot air, coupled with a swirling wind setting it all in motion. Dust devils may remain in one spot, spinning endlessly round and round until they collapse, or they can advance over the desert floor, sashaying from side to side like a belly dancer. These funnel-shaped whirlwinds can be 250 to 330 feet high (75–100 m) and from a few to greater than 300 feet (100 m) wide. The Indians of the southwestern United States believed them to be ghostly apparitions walking over the earth, possibly searching for their lost souls.

IN THE TWILIGHT ZONE

What the wind removeth, it often giveth back in the form of ridges, mounds, or seas of sand.

The wind will deposit ping-ponging sand grains when its speed decreases. Obstacles, from rocks to bushes and clumps of grass, slow the wind down, causing it to deposit the heaviest material it carries—sand. The wind dumps sand in the form of ridges and mounds called dunes.

Meeting an object it can't move, the wind whistles over and around it, leaving a pocket of slower-moving, dead air immediately behind the object. Sand grains entering this "twilight zone" slow down and are quickly pulled to the ground by gravity. As sand accumulates within the twilight zone, it becomes a growing barrier to the wind. This forces more sand to fall out, and the barrier grows.

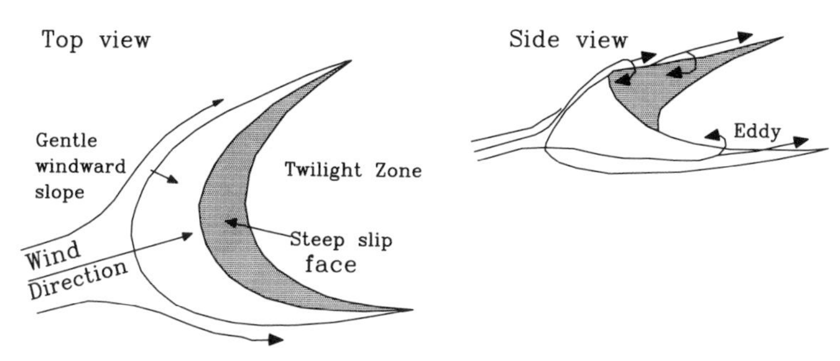

Formation of a crescent-shaped sand dune with the steep side pointing away from the wind. Sand continuously accumulates in the slow-moving air of the "twilight zone."

If there is a sufficient supply of sand, and if the wind blows steadily long enough (not usually a problem in the desert), a sand dune will form.

Sand dunes have steep sides pointing away from the wind and gentler slopes facing into the wind. Sand grains, ping-ponging up the gentle front of a growing dune, bounce merrily over the top only to find they have been dropped into the twilight zone. Stunned by this development, the grain falls to the ground and, in this manner, the back side of the dune grows steeper and steeper. Eventually it becomes too steep to support itself, and away it goes. The upper part slides down the dune under the pull of gravity. In this way the back side of the dune (called the slip face) will maintain an angle of repose between 30 and 35 degrees, the maximum slope upon which loose material will remain stable. Continued sand accumulation, coupled with sliding, permits the entire dune to migrate in the direction the wind is blowing. Dunes can hike up to 90 feet (27 m) a year in this manner.

Dunes, regardless of first impressions, are not just hopeless heaps of sand; they have their own personalities and belong to distinct families. These families form patterns that are constant and repetitive from desert to desert. Factors that affect the shapes of sand dunes within twilight zones are (1) the amount and type of vegetation present, (2) the strength of the wind, (3) the amount of

Barchan dunes, or right-handed Pilsbury doughboys, found in the St. Anthony Dunes, eastern Idaho. (Photo courtesy of R. W. Ojakangas)

available sand, and (4) the size of the blocking obstacle. Taking all of this into account, and remembering we are in the twilight zone, allows us to classify sand dunes into four types: (1) Pilsbury doughboys, which can be either left- or right-handed, (2) washboards, (3) Tootsie Rolls, and (4) stars.

Pilsbury doughboys, also called barchan and parabolic dunes, are shaped like the frozen crescent rolls that the soft, chubby fellow bakes. Right-handed ones, the normal kind, have the tips of their crescents pointed in the downwind direction. These form where the supply of sand is limited and the surface is relatively flat and lacks vegetation. They may migrate at the rate of 30–65 feet per year (10–20 m) and be as much as 1,000 feet (305 m) from flaky tip to flaky tip.

Left-handed ones, the odd kind, are similar in most respects to right-handed dunes except for the crescent tips, which point into the wind (figures, left handers are always going against the flow).

These dunes form along coasts where vegetation partially covers the sand, and the off-shore winds are strong and common.

Washboards, also called transverse dunes, are like the undulations found along dirt and gravel roads traveled by heavy trucks. Going over them your car vibrates up and down, making you feel like you're driving across a washboard. Transverse dunes are low ridges of sand separated by narrow valleys, which sit at right angles to the wind. Their spacing, as far as a camel is concerned, is about the same as the ties on railroad tracks to a walking person. These types of dunes form where there is plenty of sand and little in the way of vegetation. Many coastal dunes are of the washboard variety.

Tootsie Rolls, professionally referred to as longitudinal dunes, are long, roundish ridges of sand that parallel the wind direction. These form where sand supply is limited, the vegetation sparse, and the winds strong. Ordinary Tootsie Rolls are 10 to 13 feet high (3–4 m) and tens of feet long (>3 m). In parts of Australia and North Africa there are giant economy-size Tootsie Rolls; these babies are over 330 feet high (100 m) and 65 miles (105 km) long.

The stars of the desert are isolated hills of sand having a base that looks like Alpha Centauri. The star shape comes from armlike ridges of sand that converge upward to a central peak that can be as much as 330 feet (100 m) off the ground. Star dunes are found in areas where the wind shows no mercy, blowing in from all quarters.

Sand dunes, whether they be stars, Tootsie Rolls, or Pilsbury doughboys, range in color from brilliant white (composed of the minerals calcite, gypsum, and/or salt) to deep red, with yellows and browns in between. The red color is strictly a surface condition; it decreases with depth and increases with time. Red sand dunes are composed of the mineral quartz (that common sand mineral) covered with desert dust. Just as in the case of desert varnish, summer rains wash iron out of the dust only to have it immediately reprecipitate onto quartz grains.

Sand ripples, or minidunes, look like the wrinkled folds of skin hanging down an elephant's leg, or rolling over its broad back. Ripples are elongate, low ridges of sand that form perpendicular

A sea on which a sailor will never sail. This small sand sea is in the Algodones Dunes of southeastern California. (Photo courtesy of R.W. Ojakangas)

to wind direction, just like washboard dunes. Ripples form when gentle winds massage flat accumulations of sand. The wind has enough force to roll, pick up, and move the lighter sand grains, but the larger, heavier ones are too much for it to handle. These get left behind to form the sandy ripples.

Finally comes the seas on which a sailor will never sail a ship, seas with great waves of sand, including the great sand sea of Egypt (28,000 square miles [72,000 km^2]), the Simpson sand sea in Australia (31,000 square miles [80,000 km^2]), and the great eastern sand sea of the Sahara (74,000 square miles [192,000 km^2]), to name but a few; overall, sand seas make up about 25% of Olive Oyl deserts.

A FLASH IN THE PAN

If you are out for a Sunday drive and happen to cross a desert, you will bump over many a bridge that has nothing running

Fan-shaped piles of debris at the base of ravines and canyons are called alluvial fans. This alluvial fan lives in Death Valley. (Photo courtesy of R. W. Ojakangas)

beneath it but lizards and roadrunners. The streambeds (called washes in the United States) are dry most of the time, but when the summer rain comes—watch out.

Rain that falls in desert environments commonly does so in the form of cloudbursts—heavy downpours over a short period of time. So much rain so fast means most of it will never soak into the ground. Because the vegetation is sparse, this water flows as sheets over the surface, and down into the washes to create a horrific desert creature called a flash flood.

Flash floods come quickly, disappear even more quickly, and in between do one heck of a job of erosion. They are swift torrents of water that have the ability to transport truckloads of sand, silt, pebbles, and boulders.

Sheets of water, given a steep slope and oodles of time, carve out canyons, down which they then pour rushing walls of water. When this torrent emerges from the canyon, it immediately slows and begins to spread out, like someone opening a Japanese fan. The sand, silt, and rocks the water carries are rapidly deposited, and a fan-shaped pile of debris, called a desert, or alluvial, fan, is formed. The word *alluvial* means sediment deposited by moving water. Desert fans slope outward, away from the canyon mouth, at angles of 5–15 degrees, with coarser material deposited close to the mouth of the canyon, and the finer stuff farther away.

Alluvial fans are characteristic landforms of Mae West deserts. They have had entire cities, such as San Bernardino, California, built upon them. In the Middle East and Asia alluvial fans may be covered with small mounds of debris that represent places where shafts have been dug into them. These shafts connect with underground tunnels excavated to collect water from the upper parts of the fan. Some of these shafts and associated tunnels are more than 1,000 years old.[5]

Sinkholes—the Florida kind, with houses and cars sticking out of them—found in fans in the desert? Absolutely, and not just tiny ones either; in fact it is not uncommon for desert sinkholes to merge together, leading to the formation of ragged gullies and, over time, badlands (a land surface with little or no vegetative cover that has been intricately dissected by streams).

Sinkholes form by a process called piping. Subterranean channels are formed by water moving through unconsolidated sedimentary material. The water erodes, or washes, the sediment away, causing the upper surface to collapse and—voilà, a pipe has formed.[6] With time the pipe widens and turns into a rather splendid sinkhole.

Over the years and centuries an alluvial fan will grow larger and larger until it eventually merges with fans from adjacent canyons. When this happens a wide apron of sediment is formed called a *bajada*, Spanish for "slope."

Desert streams, those that flow for a few hours across a flat surface and those that race like a wild horse down mountain slopes to form an alluvial fan, both have their channels, or washes, sitting high above the water table. This means the water will, with time, either soak into the ground or simply evaporate into thin air. This

Heavy rain can allow a stream to flow across an alluvial fan and on, into the center of a desert basin. These streams flowed for a short time after heavy thunderstorms in Death Valley. (Photo by R. W. Ojakangas)

is the reason why most desert streams terminate in a desert basin rather than flowing on, down to the sea.

On rare occasions, when rainfall is overly abundant, streams may flow right across the alluvial fan and on, into the center of a desert basin. There they meet to form a small lake. Such lakes are shallow and short lived, lasting a few days to a few weeks, whatever is the necessary time for the water to soak into the ground or evaporate into the air. These lakes are called *playas* (Spanish for "beach"), salt pans, or just pans. After the water has gone the wet sand quickly dries out. As it loses its moisture it begins to shrink and wrinkle like a drying facial mudpack. With enough drying the surface will crack, giving the lake bed a checkerboard appearance. These "mud cracks" remain, long after the water has vanished.

The surface of playas may be coated with a fine accumulation of "salts" that have precipitated out of the water as it slowly evaporates. These salts, which give the playa the name salt pan, vary from good old table salt (halite) to a more exotic salt like sodium borate. This particular salt was mined from ancient playa lake beds in Death Valley and transported across the desert by Ronald Reagan atop a wagon pulled by a 20-mule team. Sodium borate is better known as borax.

Rivers do run through it, but only if they originate far away from a desert environment and are very, very large. The majestic Nile forms in a wet, rainy area as the Upper Nile, then flows over 1,250 miles (2,012 km) across the Sahara without the contribution of a single tributary.

Finally, the best spot in town for desert watching, other than your television set, is from great isolated hills called inselbergs. Hard, massive rocks that rise steeply from the desert floor like volcanoes from the sea, inselbergs, once upon a time, were imprisoned by softer, weaker, more abundant rocks. Erosion by wind and water, over a long period of time, removed the softer rocks and lowered the desert floor, leaving the hard, massive rock as a high-rise structure—like Ayers Rock in central Australia (1,180 feet [360 m] high)—to tower above its surroundings.

Once formed, inselbergs shed water like a duck, while the desert floor absorbs it like a sponge. In this way the desert gets lower, and the rock mass becomes higher and grander.

BLOWIN' IN THE WIND

Over time, given our knowledge of this planet of ours, it is reasonable to assume that the earth's climatic zones will shift. This happy event will occur because of topographic change, global temperature change, and the movement of jigsaw pieces from one climatic zone to the next. These various earth events can lead to the formation of deserts in areas that previously had none. Such natural changes are slow, taking hundreds of thousands to millions of years to complete. Add humans, with shovels, tractors, axes, fire, and BBQs, and the entire process can be greatly speeded up.

The advance of the St. Anthony Dunes of eastern Idaho into dry, but habitable, lands. (Photo by R. W. Ojakangas)

Desertification is not the making of cake and coffee, but the rapid, human-aided advance of deserts into nondesert areas. It is the conversion of dry, but habitable land into wasteland, due mostly to horrendous land-use policies (with a good drought or two thrown in). In the southwestern United States, because of overgrazing and overfarming, about 10% of the land area, a space equal to the size of the original 13 states, has been given back to the desert since 1900.

In semiarid lands, or steppes, which border deserts, vegetation is limited and a valuable resource. Not only does it protect the soil from drying out and being blown away, but it also feeds many different kinds of animals, and provides fuel for people. Much of the recent advance of the desert environment is due to loss of this vegetation, accompanied by its good friend drought.

In a farming area the first thing one does is to clear the land—remove the natural vegetation so crops can be planted. The natural vegetation is generally made of much sterner stuff than the crops that replaced it. If the crops fail and die, it doesn't take the wind long to come knocking at the door, demanding the mortgage payment, and there goes the topsoil.

Another effect of the loss of vegetation is the crusting over of the soil, making it relatively impermeable. This increases surface runoff, which leads to increased erosion and a decrease in the amount of arable land. It is interesting to note that on a human time scale, desertification has only occurred where people have messed with the land.

Humans have lived in grasslands for thousands of years. Western civilization was born 7,000 years ago in the fertile grasslands of Mesopotamia, between the Euphrates and Tigris rivers in southwest Asia. In fact, many of the crops we use today came from this area that is now nothing but a desert; centuries of overgrazing and overfarming destroyed the land.

You would think we could learn from our mistakes; overgrazing and overfarming of land that receives a small amount of moisture should be an absolute no-no.

The Indus River region of eastern Pakistan and northwest India is now called the Rajputana desert. Two thousand years before Christianity, it was the site of a great agricultural civilization that included two large cities, both with large granaries and brick fortresses and houses. The houses had indoor plumbing and bathrooms, and a municipal sewage system carried the wastes away. The people grew such things as wheat, barley, melons, dates, and sesame, and may have cultivated cotton. They also traded in wood, ivory, and metals.[7]

The decline and fall of this civilization was about as fast as a modern computer. It was gone by 1500 B.C. Gone because an expanding population led to an expanded agricultural base and a need to farm the land year round. Of course, the earth couldn't stand idly by and watch this without pitching in; so the planet sent a drought. Crops died, grass-cover weakened, and dust rose into the sky. As the dust storms grew the people worked the land even harder, leading to additional crop failure and more dust. Finally, much of the land was abandoned, and soon after, the cities. Today this area remains dust bowl city. It is estimated that more than 5 tons of dust hang over each square mile. During the day the sun is a hazy red, much like you see when a forest fire is burning and the sky is full of smoke.

'Tis but ancient history, say we—we of the twentieth century who know better. "The end of time arose when the black clouds rolled over the land," bringing "black blizzards." We created our own dust bowl in southeastern Colorado, northeastern New Mexico, western Kansas, and the Texas and Oklahoma panhandles; it was the 1930s and the good times were gone.

People living in the "dust bowl" area were mostly aware it was a region of drought and dust. There had been other years without rain, and other dust storms; between 1850 and 1861 there were a whole bunch of them. In fact, in 1855, dust, from what was to become the dust bowl, caused black snow to fall at Oberlin College in Ohio; from another dust storm in 1860 came black rain at Syracuse, New York. This occurred long before the grasslands were tilled over and plowed up. In fact, the area may have undergone a much more severe dry spell, one that lasted 200 years from about A.D. 1200 to 1400. This drought may have led to the demise of several ancient American civilizations, including the Anasazi cliff dwellers of the southwestern United States.[8]

Prior to 1931, most people believed that such dust storms and "drouths" would never happen again, because the way they cultivated the land would change the climate and bring more rain.

So the farmers ripped out the prairie grasses and planted wheat. Wheat was king. In 1909, in the Texas panhandle, 82,000 acres were planted in wheat; by 1927 it was 2 million.[9]

The plowing and planting of the land was furious and ferocious, for now the farmers had what was to become known as the "snubnosed monster"—the tractor. In 1915 there were 3,000 tractors in Kansas, and by 1935 there were 71,000.[10] The land didn't stand a chance.

The "drouth" began in 1932. The timing was particularly bad, for this came with the crash of the stockmarket, which drove the price of wheat down from 1 dollar to 25 cents a bushel, causing many to abandon the land, leaving it plowed, barren, dry, and open.

With or without machinery, creation of the dust bowl began the same way as it did in the Indus River valley. It began with the removal of natural grasses and flowers and the planting of wheat. Lack of rain killed the wheat, and the winds, moving sand-sized pieces of topsoil, did the rest.

> A walking man kicked up a haze as high as his waist, a running rabbit left a trail [of dust] in the air.[11]

> In the morning when you awoke in your home, the only white spot was where your head had been on the pillow.[12]

> It was widely rumored that one man, hit on the head by a drop of rain, was so overcome it took two bucketfuls of sand to revive him.[13]

It was a dust storm on August 14, 1935, that started Woody Guthrie to begin his famous song "So Long, Its Been Good to Know You." As he started to write the words he thought he was about to die.[14]

A storm on March 21, 1935, brought twilight conditions at noon to New York and New England, some 1,860 miles (2,995 km) away from the source of the dust in Colorado and Kansas.

The end of this story was a happy one; after all the suffering, hardship, and courage the rains came and the crops returned. But have we really seen the last of the dust?

The Sahel demonstrates that we have not. The Sahel borders the Sahara and crosses the entire width of Africa from Senegal to northern Sudan and Ethiopia. This was at one time an area of shrubs, clumps of grass, and small trees that receives 2.5 to 11.5 inches (100 to 300 mm) of precipitation a year. Most of this falls during a single, short rainy season.

Between 1933 and 1970 the livestock population of this area increased from 12 to 24 million. The human population also increased and so did the demand for food. In order to keep up productivity the land was not allowed to lie fallow, and animals were allowed to overgraze. Working the land this hard led to a decrease in soil fertility and eventual crop failure. The overgrazing, coupled with an increased demand for firewood, led to extensive deforestation of the area.[15]

Then, between 1970 and 1973, disaster struck. The Sahel experienced the worst drought of the century. For four successive years there was no rain; crops died, the soil dried out, and the wind moved the Sahara 95 miles (153 km) farther south. Forty percent of all the livestock perished, and famine was common and downright rampant in Ethiopia and Sudan, where more than 100,000 people died. These conditions led to a great migration south, away from the advancing desert. In 1974–1975 the rain returned and so did the people. What they found was a wasteland, an area covered by dead trees, sand-filled wells, and a bare or sand-covered earth littered with the bleached bones of cows.

Almost all the world's grasslands have been cultivated. Grass and other grassland plants are generally tough, having developed over the centuries to withstand dryness and heat. Many have developed specialized ways to beat the heat and drought. Not so your common crop planted in their place.

Will history again be repeated, or have we learned the hard lesson? Remember this: The wheat belt of the Great Plains is arid and semiarid land brought to life by constant irrigation. All of this water from pipes and ditches and wells makes it easy to forget that desert conditions are only dormant, not dead. With a few years of drought or a lowering of the water table from overuse, what is called the "breadbasket of the world" could again become part of the "Great American Desert."

Prelude to Chapter 10

Extinction: the march of scores of life-forms that come, stay awhile, then go, never to be seen again. Imagine them passing by to a heavy march rhythm that features brass alternating with the strings, punctuated by a solo trombone, loud and clear, signaling the end of yet another species. The march slowly builds in complexity and volume, finally reaching a shattering climax, which heralds a mass extinction. Then the music fades and dies out as did scores of different species during such catastrophic events.

CHAPTER 10

Leaping Lizards and Cosmic Nights

Species come, species go;
Some real fast, some real slow;
Some who say for 100 million they will stay
But end up lasting just a day;
Others are of small beginnings
And end up staying tens of millions;
But in the end it's all the same,
Eaters of plants, hunters of game,
Big or small, smart or dumb
The time will come
When a species takes its final turn
And ends up food for the burrowing worm.

How's it done, we whine and cry;
A comet falling from the sky?
Or perhaps a cosmic tilt
That gives a species the ending jilt;
On the other hand it may be ice
That makes life pay the final price.
Pick a number, take a chance,
It matters not one happenstance;

For in the end they're all the same,
Just different ways for the climate to change;
And it's this and nothing more
That opens and closes oblivion's door.

Extinction through change—earth change, life change; both are endless and full of mystery, but on vastly different time scales, like creeping continents and growing fingernails. Change is the price life pays for using the planet, and like a newborn baby, change can come painfully slowly or ride into the world full of hellfire.

Slow change comes about as close to timelessness as one can get. It seems to take an eternity to wash a mountain to the sea, to pull continents apart like pieces of taffy, or to get a seat on the earthquake ride at Universal Studios. On the other hand, hellfire change is much more like a car accident; it comes suddenly and unexpectedly and may well be catastrophic. The blink of an eye for a hurricane to level a town, the movement of the second hand for a mountain to explode and bury the countryside, and just a heartbeat to get pregnant.

In professional football it is the job of a defensive end to disrupt plays and tackle quarterbacks. In the physical world this position is held by climatic change, whose job it is to disrupt weather patterns and throw life for huge losses. Wrecking havoc on weather patterns can bring short periods of terrible weather (like the cold, snowy winter of 1994), eons of ice and snow, or ages of dusty hot drought.

Fortunately for the earth (and life) climatic change is just that—change. Bad weather eventually fades like an old flame; the earth warms and the ice and snow melt, or the rains come and the hot dust is covered by grass. To the earth it is just another afternoon at the movies, and when the show is over, the planet simply "gets on with it." But what about life? Drastic climatic changes are of extreme importance to life forms that live on, or near, the planet's surface, and for them, there may be no "getting on with it." Much like the "terminator" of movie fame, climatic change can represent the end of the road—oblivion for an entire species.

Over a period of some 570 million years hordes of different species have checked into a resort called Earth, sun-tanned or snorkeled for awhile, then vanished forever. The local (or geographically restricted) extinction of species is as common as penalties in a hockey game. A list of the fallen would stretch from Casablanca to the Magic Kingdom. Some of North America's better-known ex-guests include the trilobite, ejected for hating cold water, the wooly mammoth and saber-toothed tiger, both gone for an eternity trying to outrun the ice and cold, and the great sloth but a page in history because of bad snacking habits (actually this creature survived by slowly evolving into a couch potato).

In the harsh world of everyday life, local climate changes can mean everything from discomfort to death; be it cold, drought, snow, flood, heat, or climate-induced migration, the environment has changed and life either adapts or dies—and that's it folks. Of all the species that have ever existed on this planet, the vast majority are no longer with us.[1]

So local comings and goings are common. What is rarer than a pep rally for a science club is a mass slaughter, a complete elimination of hundreds of different species over a relatively short time period—wholesale emigration to never-never land. Since the start of the Cambrian period (about 570 million years ago) there have been fewer than 10 mass extinction events documented in the fossil record. Mass extinctions must be the earth's version of the Olympic Games. Once every 60 million years all the species gather, and when the games are over it's not gold or silver for the winners, but the chance to keep on living—and don't bother to bring your American Express card or your Visa, the earth takes neither.

So every once in awhile something terrifically terrible takes place on the surface of the planet. Unfortunately, we who now live have no certain answers as to what this terrible event may be. There were no CNN reporters at the frontlines, no television movie makers around to offer the trilobite a million or two for the story of its last gasp. There was no one to cover some of the greatest stories of all time—the great Cambrian catastrophe, the Devonian demise, the Permian purge, and the Ordovician death orgy, to name just a few.

The truth of what caused these mass extinctions may never be known for certain. So, amongst all the speculation, the theories and the arm waving, we continue to live on the edge; with our American Express cards clutched in our hands we wait for the comet to fall, the ice to come, the volcano to erupt, the ________ (feel free to fill in the blank; after all, it's your planet too).

A TIME TO DIE

"It's a good day to die," or so the movie line goes. For creatures like trilobites and dinosaurs it would be much more accurate to say, "it was a good millennium in which to become extinct." To add insult to oblivion the dinosaur and its friends didn't fail at adapting to local changes. They were victims of a worldwide, catastrophic environmental change that they had not a hope in a wicker basket of adapting to, and that's a real bummer because it could happen to any one of us (speaking species-wise).

Here you are, king of the swamps, queen of the flowering plants and grasslands, and no matter what the earth or climatic change throws at you, you manage to adjust, to survive, even to flourish for more than 150 million years. Then, almost overnight, the sky darkens, the wind turns bitter, and something wicked this way comes. A catastrophe, a global disaster, and what took 150 million years to create, refine, and perfect is gone in less than 10 million (we who watch TV and eat off plates have been here for about 4 million years). Not only did this disaster sweep away the king and the queen, but it also took most of their court with them—flying reptiles, plesiosaurs, ichthyosaurs; over one-half of all species alive were gone. This was the end of the Cretaceous period (65 million years ago), a great time for fossil making. It was also a better time to be just about anywhere else—even Fairbanks in January would be wonderful by comparison.

Fossil making has been an earth sport since the end of the Cambrian, about 570 million years ago. At this time nothing walked on land; everything either swam, floated, or lay in the muck, so fossil making was restricted to oceans, lakes, rivers, and mud puddles. This was also true during the Ordovician debacle (440–450 million years ago) and the Devonian rout (360–370 million years ago). During these times of fossil making the rock record tells a story of a cold earth, of great ice sheets covering large parts of the continents, and of drastic lowering of sea levels.

Three times over a 140-million-year period the earth went cold, and three times life in the seas went faster than lobsters in a pot. Cold water and lowering of sea levels are not what brings joy and happiness to the heart of marine creatures like trilobites, brachiopods, or ammonites, to name but a few. In fact, the vast majority of the species that became extinct during these times lived in relatively warm water. Similar creatures living in the colder, northern climes survived with just a touch of frostbite.

The end of the Permian period, some 250–255 million years ago, marked the first time land animals were put up for fossilization along with marine creatures. This particular mass extinction is said to have been the greatest of all time. It is estimated that 80% of marine creatures were wiped out, along with 75% of amphibian families, over a 5- to 10-million-year period.[2] Plants were also decimated in numbers and varieties. It was also about 250 million years ago that the continents smashed together to form the supercontinent of Pangaea. Part of this great landmass sat over the South Pole, and you can guess the rest. It was natural refrigeration for the land mass. Once again the earth cooled, sea levels fell, the ice came, and life went out the back door. It was also at this time that some 1.3 million cubic miles of lava flowed across northern Siberia to produce what is called the Siberian Traps (in geology a "trap" is defined as any dark, fine-grained, igneous rock). Perhaps debris from these eruptions blocked incoming sunlight for a sufficient length of time to trigger an ice age.

After each of these "great dyings" the miracle that is life managed to rebound. But, for life, the rebound takes millions of years, and in the case of the great Permian bake-off, life managed only a small bounce, reaching about half of pre-Permian levels. Mass extinction seems to permanently reduce species, decrease earth population, and give the underdog a chance to have its day. The dinosaurs were given their big opportunity some 200 to 215 million years ago (at the end of the Triassic period) when the supercontinent of Pangaea broke apart. Once again there was a mass extinction and the reigning royal family was wiped out. These kings and queens were great, crocodilelike creatures (thecodonts) who would have slurped up Captain Hook, the alarm clock, and the first mate, Smee, and still had room for a snack or two before breakfast.

With the large land animals going extinct, the ancestors of the dinosaur scurried in to fill the ecological void. They did this so well that they ended up starting a dynasty that lasted for more than 150 million years. In fact, dinosaurs may have been far more successful than we can even imagine; they may have pulled a real Houdini and slipped right through the Cretaceous disaster and on into the future. According to paleontologist Jack Horner, they turned themselves into birds and flew away to evolve into our Thanksgiving Day treat (maybe they weren't so smart after all).[3] As farfetched as this may sound, many scientists believe dinosaurs really did evolve into birds. They certainly share many similarities, from being warm-blooded and having the same bone structure to showing similar nesting behavior.

Speaking of creatures that step in to fill ecological spaces left by dying species, I would like to point out that 9 times out of 10 these niche fillers are animals with fast reproductive rates, and almost no scruples as to what they will eat. If the larger mammals of today were to bite the dust—the tigers, elephants, rhinos, hippos, whales, apes, zebras, cows, and humans—the world would end up being run by flies, rats, roaches, and weeds.

Dinosaurs ran the earth for 150 million years—that's a very long time. To survive that long, dinosaurs had to be extremely resourceful, adaptable, and lucky.

I can remember the days when dinosaurs were portrayed as cold-blooded, stupid beasts who plodded along dragging their enormous tails behind them. Remember the classic pictures of the huge dinosaur head and, there, right beside it, a drawing of a walnut, which was supposed to depict the size of the animal's brain. It turns out that the walnuts belonged in the heads of the paleontologists who painted that picture. As we all now know, dinosaurs were fast, smart, cunning, and maternal. They were also extremely adaptable, being able to survive from arctic climates right down to the equator.

However, their adaptability didn't help them when the planet began to change big time about 75 million years ago. The dinosaurs, according to many paleontologists, started to die out some 10 million years before the so-called Cretaceous event—the coming of the great asteroid. From 75 to 65 million years ago dinosaur species decreased by 70%; over a 10-million-year period, the number of dinosaur genera (any group of similar things—like stooges, politicians, or pigeons) went from 30 to 13, based on fossil evidence from Wyoming and Alberta.[4] During the same 10 million years those fuzzy pests called mammals were rapidly diversifying, and becoming much more abundant and daring. Instead of sneaking about only in the middle of the night, some were taking to showing their pointed snouts during the day. It would certainly appear that the dinosaurs were following the normal route of extinction; they were going with a long, echoing whimper and not a big bang. But what of the great asteroid? That cataclysmic impact that humbled the earth? Well, the apparent slow changing of the guard does not necessarily negate the asteroid's importance in finishing off the dinosaurs. As Ken Hsu has pointed out, "a bomb on an old folks' home is still a major disaster."[5]

The Cretaceous extinction, the destruction of the dinosaurs and their friends, was not the last great extinction to occur on the planet. There is one going on right now. We humans are killing one species every few days; that's a rate about 400 times faster than the rate of natural extinction. To make room for us, and our rabbits, cats, cows, sheep, dogs, gerbils, and late-night talk-show hosts, we are carrying out another "great dying." Ten thousand years ago there were 5 million people on this planet, in 1850 about 1 billion, in 1950 2.5 billion, and today there are 5.6 billion! This earthwide, human population expansion is a terrible threat to the biosphere; from the rain forests to elephants, whales, and ozone, we are simply killing this place. A species here, a species there; so what and who cares? We should. Have you seen the movie *Jurassic Park*? If so, do you remember the part where the mathematician is trying to explain the theory of chaos? "A butterfly flaps its wings in China," he says, "causing a rain storm in Central Park." Tiny ripples build upto a major catastrophe. A species here, a species there, the condor and spotted owl, both are small ripples in the pool of life. Small ripples that spread to one day become giant waves that come ashore in unexpected places.

Remember the children's poem:

For want of a nail
The shoe was lost,
For want of a shoe
The horse was lost,
For want of a horse
The rider was lost,
For want of a rider
The battle was lost,
For want of a battle
The kingdom was lost,
And all for the want
Of a horseshoe nail.
—Anonymous

DINOSAURS TO PINATUBO

Dinosaurs, ice ages, trilobites, falling meteors, great sea turtles, and Mt. Pinatubo are all linked together by that which we love to talk about but can do nothing to change—the weather, or in the long-term mode, climatic change. Ice ages come and glaciers form because the climate cools, volcanoes erupt and asteroids fall, and the climate on the surface can take a nose dive or get a high fever; dinosaurs, sea turtles, and trilobites all depend on a relatively constant climate to survive and flourish and plan picnics and ski trips.

Climatic change is a quick and easy way to get the entire planet to warm up or cool down, either of which can be bad news for scores of life forms. Temperature change permeates the earth's ecosystem on a global scale, and mass extinctions are definitely international events. Temperature is the primary limiting factor controlling the geographic distribution of life, and cooling or warming of the tropical areas represents the one common thread tying mass extinctions together.

Over 570 million years and through each and every mass extinction the hardest-hit species are those that live in tropical regions (and you wanted to move south). As one example among many, at the end of the Triassic period (about 200 million years ago) burrowing bivalves (mollusks, like oysters and clams, which have a shell that consists of two parts connected by a hinge) were virtually wiped out in the southern parts of North America. However, 900 miles (1,450 km) to the north, in what is now North Dakota and Manitoba, similar creatures continued to flourish as if nothing had occurred. What this means is that global cooling can easily disrupt and eliminate the tropical climatic zone, but it has a much harder time trying to bump off the earth's other climatic areas, the temperate and polar zones. As the earth cools, species living in more northerly climes can pack up and travel south (like Minnesotans do in February); for life in the tropics there is just no place to move to. If global warming takes place, just the opposite occurs; the polar regions melt away and species in the tropics have the option of taking a northern vacation. Because the geologic record shows that most mass extinctions have hit the tropics the hardest, it seems reasonable to assume that most have occurred because of global cooling.

Finally, one last link between climate and mass death. Fossil evidence shows that extinctions do not occur all at once; the planet does not just go frigid and all species in the tropics turn into ice cubes. Instead, what appears to happen is that different species become extinct at slightly different times. This can be explained by a slow temperature change that causes the so-called thermal threshold (sort of like the rating on sleeping bags, only these are on life-forms) of different species to be reached at different times as the earth cools down or warms up. For instance, the Floridian might go when the temperature hits 45° F (7° C), the Kentuckian at 32° F (0° C) whereas the Minnesotan will hang in there all the way to -30° F (-34° C).

Climatic change and slow cooling of the planet to trigger mass extinctions bring us directly to the terrible lizards. Did they go over several million years due to global cooling? Did they die in a violent flash of flame, smoke, and dust? Or did they get it both ways, the latter finishing off what the former had begun?

The most recent controversy on the demise of the dinosaurs started in 1980 when Louis and Walter Alvarez (and others) proposed that the dinosaurs died in the cold and dark when a giant asteroid or comet smashed into the earth (a vast swarm of shooting stars entering the earth's atmosphere and vaporizing as a group would produce the same effect). Such an impact could cause continent-wide fires and throw zillions of dust bits into the atmosphere. The dust, smoke, and pulverized asteroid stuff would dramatically decrease the amount of sunlight reaching earth. Global cooling would take place, or as the more dire-hearted envision, nuclear winter would stalk the landscape. Nuclear winter would cause the flowering plants to wither and die; the herbivorous dinosaurs (those giant plant recyclers), with nothing to eat, would die by the herdful, soon followed by the meat eaters. And the dinosaurs wouldn't go alone; others to join them would include flying reptiles and scores of marine life-forms from floating algae to ammonites, belemnites, and large marine reptiles. In the real "Cretaceous tragedy" over 50% of all species living became toast.

This theory of an "asteroid impact" has become very popular, more so since Comet Shoemaker–Levy 9 smashed into Jupiter, demonstrating to one and all that it can actually happen. Evidence for the earthly impact is provided by the presence of a thin layer of clay (called the boundary clay layer) that occurs at the break between the Cretaceous period and the Tertiary period—the time transition between dino in and dino out. This clay layer has been found in the "right place" at more than 70 sites around the world, and everywhere has what many scientists consider the "right stuff": it is abnormally enriched in a rare, metallic element called iridium.

A metal of the platinum group of metals, iridium is rare (0.1 parts per billion) in the earth's crust. However, iridium is abundant in meteorites and, therefore, the anomalous amounts in the clay layer (4–7 parts per billion) could be explained by fallout of pulverized meteorite dust over the entire planet, following an explosive impact on a large scale (it has been estimated that the asteroid would have had to have been more than 6 miles in diameter). This iridium-rich clay layer has been age dated at 65 million years; thus it formed at the same time that many species were going extinct.

What makes this theory really exciting is that we humans have always had a great fascination for things from outer space that are trying to bump us off—thunderbolts of the gods, UFOs, meteorites and asteroids, and H. G Wells's "War of the Worlds," which caused panic when first read over the radio in 1938. The fascination with dinosaurs being paid a visit by an asteroid is based on global evidence, the popularity of dinosaurs, the Jupiter event, and the power of imagination—a world in flames and smoke, dark and cold, a nuclear winter causing mass starvation and death. It certainly appeals to our dark side, that side of us that walks hand-in-hand with the unknown and unexplained.

However, there are those who would disagree, including paleontologists who have worked for years on dinosaur fossils and geologists who have problems with the "unexplained" side of the impact theory. Some of the more interesting "anti" tidbits are summarized below.

1. According to the impact theory, the power of the asteroid impact (and thus the amount of dust tossed into the stratosphere) was 1,000 times greater than that of the eruption of Krakatoa. We know that volcanic ash from Krakatoa remained in the upper

atmosphere for a 2- to 3-year period, which coincided with global cooling. However, 65,000 years ago the volcano Toba erupted on the island of Sumatra. This eruption was 400 times greater than that of Krakatoa, but nothing much happened—no extinctions, no apparent global changes, not a single game canceled. If an asteroid impact 1,000 times greater than Krakatoa wiped out half of the species on earth, then a volcanic eruption 400 times that of Krakatoa should have gotten at least a few of them.

2. It has been written that the iridium anomaly in the clay layer must be due to an impact event because (a) iridium is rare in the crust, (b) it is rare in material erupted from volcanoes, and (c) it is common in meteorites. But is this really true? Or is the old saying "seek and ye shall find" more appropriate? In 1983 it was discovered that volcanic ash collected over Kilauea was enriched in iridium by a factor 10,000 times that of normal Hawaiian basalt.[6] Volcanic dust found in Antarctic drill cores is enriched in iridium with values ranging from 4–7 parts per billion, similar to the values found in the Cretaceous clay.[7] A volcano on Reunion Island (another hot spot volcano like Kilauea) has produced volcanic ash with iridium values of 4–8 parts per billion, and from Krakatoa and Pinatubo we know volcanic ash can be deposited worldwide in a matter of a few days. On the meteor side of the ball, iridium anomalies from other impact sites range from 0.55–4 parts per billion, less than in the boundary layer, and in some of earth's own volcanoes.

3. When many of the Cretaceous–Tertiary boundary sections are examined in detail, they do not show a sudden "iridium peak," that is, a sudden jump from 0.1 parts per billion iridium to 4–7 parts per billion (impact) then back to 0.1 parts per billion again, which you would expect from an event that suddenly puts tons of iridium-rich dust into the air. Instead, the clay layers show a gradual buildup from 0.1 to 4–8 parts per billion, and then a gradational decrease back to 0.1 parts per billion, occurring over the time interval it took to form the entire clay layer—some 100,000 years.[8] This pattern is much more in accord with large, continuous volcanic events. For example, the Deccan Trap basalts in India show a gradual increase of iridium, a peak, then a gradual decrease over a

200,000-year time span (you might expect a huge volcanic eruption like a Yellowstone to give an iridium signature closer to that of an impact event).

4. Not only do we have anomalous iridium in the boundary clay but we also have anomalous gold, copper, osmium (another platinum group element), chromium, nickel, arsenic, and molybdenum.[9] Most of these elements are more characteristic of the earth and its volcanoes than of outer space and meteorites.

5. A huge asteroid impact that swept away dinosaurs and ammonites should also have done in almost any species that does not adapt quickly to environmental change—such as frogs and turtles; they were there then and they are here now, and they never missed a beat.[10] These creatures, however, are adaptable to slow climatic change, and a cold earth wouldn't phase them one bit.

6. The collision between Shoemaker–Levy 9 and Jupiter proved what most people already knew: Comets and asteroids do, on rare occasions, collide with planets, the collisions are violent and powerful, and the earth has, in the past, been an asteroid dartboard. What was amazing, astounding even, was the power and violence of the Jupiter smacking. "It was like God striking the planet" said astronomer Glen Orton. One of the comet's fragments, "G," hit the planet with a force 100,000 times the power of the largest nuclear bomb ever exploded on earth. According to one account this would be like stacking 8 pounds of TNT on each square foot of land on earth and then blowing it up.[11] These numbers are certainly mind-boggling, but what they really mean is that our vision of what a so-called nuclear winter or asteroid sunset would be like may be terribly underexaggerated. Such an impact and explosion on earth, by a fragment much smaller then the size of the one that supposedly did the dinosaurs in, would most likely have wiped out every living thing on the planet. Life would have been back to square one, right at the bottom of the sea, which would be the safest place on the planet. Do you think such an impact would have spared any of the mammals? Make up your own mind; after all, it is all conjecture and I sure hope it stays that way.

The cosmic end of the dinosaurs.

7. As already mentioned, fossil evidence shows that dinosaurs were very much on the decline long before the great event occurred. Something other than comets from space appears to have been taking place; something long-term, something cold, and something of vital importance to the dinosaur. Fossil evidence suggests that dinosaurs were taking the same slow road to extinction that many other species had already traveled. If an asteroid did hit the earth it may simply have put the finishing touch on an "old dinosaurs'" home, one that already had many "room for rent" signs in the windows.

The meteorite impact theory is but one of many that have been proposed to explain the death of the dinosaurs. Some the others are funny, like a diet change gave the dinosaurs chronic constipation, and in agony and all-a-bloat, they perished. Or the one that claims they grew so big they could no longer mate. Then there are the silly ones: Dinosaurs died from lack of oxygen, they couldn't fit on Noah's ark, or they were taken off the planet by aliens—imagine the earth as a giant dinosaur ranch, tended by four-armed cow-creatures from Gamma II. Finally, there are the explanations that fall within the realm of possibility, such as one involving volcanic eruptions.

Volcanic eruptions have been linked to climatic change from the time of Ben Franklin to now. As we have previously seen, eruptions in Iceland, Indonesia, and the Philippines have caused global cooling, but with these more recent eruptions the cooling lasted only a couple of years at the most. Could a great volcanic eruption, such as a Yellowstone, a Mt. Mazama, or a long series of closely spaced eruptions, such as the 200,000-year activity to form the Deccan Traps, put enough ash and gas into the atmosphere over a long enough period of time to trigger a long-term climatic shift? We just don't know.

Cores from the deep-sea drilling project show that the amount of volcanic ash in sediments on the seafloor increased dramatically about 2 million years ago and has stayed relatively high since then. This coincides nicely with the start and duration of the last ice age.

The 1963 eruption of Bali's Mt. Agung, which Indonesians call the belly button of the world, shot gas and ash high into the stratosphere (>6 miles [10 km]). This material quickly encompassed the globe and caused red sunsets, halos around the moon and sun, and an increase of the earth's upper stratospheric temperature by 110° F (6° C) for a 3-year period. During this time the surface temperature decreased by 0.75° F (0.4° C).[12] Now, in terms of volcanic eruptions, the old belly button wasn't much; the eruption of Krakatoa was 10 times greater, Tambora 1,000 times greater, and a Yellowstone? If a volcano could put enough ash, sulfur, and other stuff into the stratosphere to decrease the surface temperature by 4° to 8° F (2.2°–4.4° C) over a relatively long period, then it could well trigger permanent global cooling, and the start of an ice age. This would be especially true if there was a large landmass over one of the poles at the time of the eruption or eruptions.

Some scientists have called upon the volcanic activity that formed the Deccan Trap basalts, which occurred about 65 million years ago, as the culprit in the demise of the dinosaur. However, what's true for meteorites is also true for volcanoes. Dinosaurs seem to have been going extinct well before this volcanic activity started. However, the eruptions did occur over a 200,000-year period and may have been the straw that broke the dino's back. The hypothesis of Ken Hsu (1986) that cosmic impacts caused "darkness at noon" could just as well be ascribed to big volcanic eruptions; just ask the people of Yakima, Washington, who had 600,000 tons of ash dropped on them one spring morning in 1980.

Finally, before we are through with dinosaurs, brief mention needs to be made of the forgotten life-forms that were the base of the dinosaurs' food chain—the plants. During the late Cretaceous, seed-bearing plants dominated the landscape; these were flowering plants and conifers. At the end of the Cretaceous these plant types were largely replaced, for a short 2- to 3-million-year period, by a flora dominated by ferns. Ferns are among the first plants to colonize the barren surface of newly formed and cooled lava. Ferns are excellent invaders, but over time they lose out to seed-bearing plants, which are better competitors for sunlight, space, and nutrients.

The event at the Cretaceous–Tertiary boundary killed off the higher plants much more severely in the southern climes than in the north. Three-quarters of all species living in what is now New Mexico, half of those in Wyoming, and a quarter of those in central Alberta went down. At this time New Mexico had a tropical climate, Wyoming was temperate, and Alberta was a bit colder.[13] A global cooling event would have been more disastrous to plants living in tropical regions compared to those already flourishing under colder conditions. But a sudden impact of large proportions, causing "darkness at noon," would have wiped them all out.

Flowering plants and dinosaurs seem to have gone together over a 10-million-year period of time, with plants leading the way. Once the tasty fast-food plants were gone, the dinosaurs must have figured enough was enough and went with them—I mean, would you eat ferns?

THE FINAL COUNT

Over 1,000 years ago the inhabitants of northern Europe well understood the effect of climatic change on life. Many of their folk tales and legends were built around fierce storms and long periods of bad weather. One such tale, "Finbul Winter," tells of a cold earth and a long period of ice and snow: "Heavy snows are driven and fall from the world's four corners; the murdering frost prevails. The sun is darkened at noon [Hm, this sounds much like recent guesses as to what nuclear winter or an asteroid impact may have brought]; it sheds no gladness; devouring tempests bellow and never end. In vain do men avail the coming of summer."

Climatic change can certainly devastate life, but as the late Derek Ager wrote, "the whole history of the earth is one of short, sudden happenings [such as Finbul Winter] with nothing much in particular in between." Fortunately, for the less adventurous of us, we seem to inhabit a relatively lazy, let-things-go-as-they-are planet. It was Teddy Roosevelt who said, "speak softly and carry a big stick." The earth seems to speak softly to life for a millennium; then, out of the blue, down comes the big stick. Its purpose is not to slap a hand or a paw, but to create a full-blown, hellfire catastrophe.

The earth, to some people, is a dynamic organism where everything is related to everything else, like James Lovelock preaches in the concept of Gaia. This is similar to the system of checks and balances in a democratic government, set up largely to prevent excesses by one branch (or person) over another. If this concept has any validity in the physical world, then humanity's excesses, our abuse of the biosphere and our lack of respect for other living things will eventually, as T. S. Eliot said, lead "man down into the pit and all his thoughts will perish." If this happens then possibly humanity's only monuments to itself will be "the asphalt road and 1000 lost golf balls." If Elliot had written this today he probably would have added plastic bottles and gummy worms.

Personally, I like the idea of a dynamic earth, a planet that takes care of its own, one way or the other. This of course means that there is no need to call on space projectiles and creatures from other worlds to clean the planet and help life on its way. The earth can manage this just fine, thank you very much.

The really good side of living on a dynamic planet is that we are part of the orchestra, and, as such, we do have some say in how the music will play. Oh, I know, nature is still the conductor. We can't stop the continents from creeping, volcanoes from exploding, earthquakes from rumbling, or the climate from changing (at least not yet). But we can make an attempt to see that the planet stays on the straight and narrow. We can do our best to keep the air, water, and land in as "prime earth condition" as possible. We can help other species survive our onslaught, and we can make a real effort to contain the spread of our own kind. This of course means that we must try to live within the earth's means—a natural budget for all of us.

THE MUSIC PLAYS ON

So, what goes around comes around. The music of the earth plays on, repeating itself over and over. From the solo performances of volcanoes, earthquakes, glacial ice, rolling waves, blowing wind, and hot springs to the grand symphony of creeping continents, moving oceans, and changing seasons, earth music can be beautiful, awesome, and inspiring, or unsettling, terrifying, bone rattling, stomach churning, and downright deadly. But no matter what music the earth plays, no matter what we have to put up with from the planet, we must always remember that this planet has given us the greatest gifts of all: land, air, water, and life.

> Touch the earth, honor the earth, her plains, her valleys, her hills and her seas; rest your spirit in her solitary places.
>
> —Henry Beston

The music of the earth and life; you can't have the one without the other. So, as we ride along on the corkscrew of time, we need to pay close attention to, and try to understand, the music of the earth, which is, when all is said and done, truly a celebration of everything that is called life.

References

CHAPTER 1: Those Creepy Continents

1. Alfred Wegener, *The Origin of Continents and Oceans* (London: Methuen, 1924).
2. Jonathan Weiner, *Planet Earth* (New York: Bantam Books, 1986), 37.
3. A. Hallan, *Great Geological Controversies* (Oxford: Oxford University Press, 1983), 131–133.
4. Carla Montgomery and David Danthe, *Earth: Then and Now* (New York: Wm. C. Brown, 1991), 135–138.
5. Quoted in Weiner, *Planet Earth*, 18.

CHAPTER 2: The Fire Within

1. Robert Decker and Barbara Decker, *Volcanoes* (New York: W. H. Freeman, 1989), 172.
2. Leonard Palmer, *Mt. St. Helens: The Volcano Explodes* (Seattle: Caroline House Publishers, 1980), 9.
3. Derek Ager quoted in Jonathan Weiner, *Planet Earth* (New York: Bantam Books, 1986), 198.

4. Howel Williams, *Crater Lake: The Story of Its Origin* (Berkeley: University of California Press, 1963), 29.
5. Tom Simkin and Richard Fiske, *Krakatau 1883: The Volcanic Eruption and Its Effects* (Washington, DC: Smithsonian Institute, 1983), 155–158.
6. Ibid.

CHAPTER 3: Living on an Eggshell

1. "Villagers Search for Survivors after Deadly Quake in Japan," Associated Press, July 14, 1993.
2. "Quake Toll Tops 5,000; Stocks Dip," Associated Press, January 24, 1995.

CHAPTER 4: And Away We Go

1. R. G. McConnell and R. W. Brock, *Report on the Great Landslide at Frank, Alberta* (Ottawa: Canadian Department of the Interior, Annual Report, 1902–1903), 7.
2. Ibid.
3. Stephen Harris, *Fire and Ice: The Cascade Volcanoes* (Seattle: The Mountaineers, 1980), 213–219.
4. Ann McGovern, *Stone Soup* (New York: Scholastic, 1968).

CHAPTER 5: When The Waters Dance

1. Jonathan Weiner, *Planet Earth* (New York: Bantam Books, 1986), 292.
2. Ibid.
3. Gordon McDonald, *Volcanoes* (Englewood Cliffs, NJ: Prentice-Hall, 1983), 328.
4. Ibid.
5. Robert Decker and Barbara Decker, *Volcanoes* (New York: W. H. Freeman, 1989), 210.
6. Fredrick Lutgens and Edward Tarbuck, *Essentials of Geology* (Englewood Cliffs, NJ: Prentice-Hall, 1995), 214.

7. Rachel Carson, *The Sea Around Us, Revised Edition* (New York: New American Library, 1961), 62.
8. J. M. Nash, "How Did Life Begin," *Time*, October 11, 1993, 68–74.

CHAPTER 6: Harvesting the Earth

1. "Pyramids," *World Book Encyclopedia* (Chicago: World Book, 1990), 921.
2. George Goodwin quoted in F. D. Adams, *The Birth and Development of the Geological Sciences* (New York: Dover Publications, 1938), 283.
3. Adams, *Geological Sciences*, 282.
4. James Craig, Robert Vaughan, and Brian Skinner, *Resources of the Earth* (Englewood Cliffs, NJ: Prentice-Hall, 1988), 199–201.
5. Carla Montgomery and David Dathe, *Earth: Then and Now* (New York: Wm. C. Brown, 1991), 493–494.
6. *Mineral Commodity Summary* (Washington, DC: U.S. Bureau of Mines, 1994).
7. Ibid.
8. Dixie Lee Ray and Lou Guzzo, quoted in Mineral Information Institute, Update #11, Denver, 5.
9. Craig *et al.*, *Resources*, 228.
10. Pierre Berton, *The Klondike Fever* (New York: Alfred A. Knopf, 1958), 47.
11. Murry Morgan, *One Man's Gold Rush* (Seattle: University of Washington Press, 1967), 112.
12. Ibid, 117.
13. Ibid, 90.
14. Pierre Berton, *Klondike Fever*, 60.
15. Ibid, 82.
16. David Cleary, *Anatomy of the Amazon Gold Rush* (Ames, IA: University of Iowa Press, 1990), 1–6.
17. Ibid, 20–21.
18. Victor Argenzio, *Diamonds Eternal* (New York: David McKay, 1974), 13.

CHAPTER 7: Winter's Breath

1. Mark Twain, *A Tramp Abroad* (New York: Harper and Brothers, 1907), 141–151.

2. Richard Ojakangas and Charles Match, *Minnesota's Geology* (Minneapolis: University of Minnesota Press, 1982), 114.
3. Ibid, 102.
4. S. Chernicoff and R. Venkatakrishnan, *Geology* (New York: Worth, 1995), 497.
5. Ibid, 559.
6. "Ancient Tree Rings Show No Evidence of Global Warming," Associated Press, May 18, 1994.

CHAPTER 8: Life on the Edge

1. A. Duxbury and A. Duxbury, *World's Oceans*, 4th edition (New York: Wm. C. Brown, 1994), 239.
2. Steven Porter and Brian Skinner, *The Dynamic Earth* (New York: John Wiley and Sons, 1989), 294.
3. Ibid.
4. H. Wenzel, *The Eternal Sea* (London: Abeland-Schumann, 1979), 20–21.
5. P. Groen, *Waters of the Sea* (London: D. Van Norstrand, 1967), 186.
6. H. Wendt, *The Romance of Water* (New York: Hill and Wang, 1969), 50–52.
7. Frederick Lutgens and Edward Tarbuck, *Essentials of Geology* (Englewood Cliffs, NJ: Prentice Hall, 1995), 275–276.
8. Ibid, 276.
9. Duxbury and Duxbury, *World's Oceans*, 257–258.
10. Ibid, 260.
11. Ibid, 319.

CHAPTER 9: Summer's Fire

1. Stewart Douglas, quoted in M. A. J. Williams, "Cenozoic Evolution of Arid Australia," in H. G. Cogger and E. E. Cameron, eds., *Arid Australia* (Sydney: Australian Museum, 1984), 59.
2. Steven Porter and Brian Skinner, *The Dynamic Earth* (New York: John Wiley and Sons, 1989), 242.
3. Edmund Jaeger, *The North American Deserts* (Stanford: Stanford University Press, 1967), 128.

4. Ron Cooke, Andrew Warren, and Arthur Goudie, *Desert Geomorphology*, (London: University College London Press, 1993), 46.
5. Porter and Skinner, *Dynamic Earth*, 245.
6. Cooke *et al.*, *Desert Geomorphology*, 83–84.
7. Reid Bryson and T. Murray, *Climates of Hunger* (Madison: University of Wisconsin Press, 1977), 107–114.
8. "U.S. Saw Drought during Middle Ages," *Washington Post*, June 22, 1994.
9. Douglas Hurt, *The Dust Bowl: An Agricultural and Social History* (Chicago: Nelson Hall, 1981), 25.
10. Paul Bonnifield, *The Dust Bowl* (Albuquerque: University of New Mexico Press, 1979), 49.
11. Michail Pafit, "The Dust Bowlers," *Smithsonian*, June, 1989), 49.
12. Ibid.
13. Ibid.
14. Bonnifield, *The Dust Bowl*, 2.
15. John Stewert and Holm Tiessen, "Grasslands into Deserts," in Constance Mungalt and Digby McLaren, eds., *Planet under Stress* (Oxford: Oxford University Press, 1991), 199–201.

CHAPTER 10: Leaping Lizards and Cosmic Nights

1. Steven Stanley, *Extinction* (New York: Scientific American Books, 1987), 1.
2. Ibid, 19.
3. William Allman, "Hunting for Dinosaurs", *U.S. News & World Report*, June 7, 1993, 72.
4. Stanley, *Extinction*, 160.
5. Derek Ager, *The New Catastrophism* (New York: Cambridge University Press, 1993), 186.
6. C. Officer, "Extinctions, Iridium, and Shocked Minerals Associated with the Cretaceous–Tertiary Transition," *Journal of Geological Education*, 38, 1990, 409.
7. Ibid.
8. Ibid, 409–410.
9. Ager, *Catastrophism*, 185.
10. Michael Lemonick, "Rewriting the Book on Dinosaurs," *Time*, April 26, 1993, 49.

11. Martin Merzer, "Fireworks from Jupiter," *Knight-Rider Newspapers*, June 26, 1994.
12. Robert Decker and Barbara Decker, *Volcanoes* (New York: W. H. Freeman, 1989), 217.
13. Stanley, *Extinction*, 157–160.

Bibliography

The idea for adding the music to the bibliography came from a 1988 paper by Charles Carlton Plummer, "Music to Soothe the Savage Physical-Geology Student," in the *Journal of Geological Education* 36, 88–89.

General: The whole earth orchestra playing Paul Winter's "Voices of the Planet."

Adams, F. D., 1938. *The Birth and Development of the Geological Sciences*. New York: Dover.

Ballard, R. T., 1983. *Exploring Our Living Planet*. Washington, DC: National Geographic Society.

Bates, R. L., and Jackson, J. A., eds., 1984. *Dictionary of Geological Terms*. Garden City, NY: Anchor Press/Doubleday.

Dixon, D., ed., 1989. *The Planet Earth*. Chicago: The World Book Encyclopedia of Science.

Hazen, R. M., 1982. *The Poetry of Geology*. London: Allen and Unwin.

Lutgens, F. K., and Tarbuck, E. J., 1995. *Essentials of Geology*, fifth edition. Columbus, OH: Merill Publishing. Chapters 8, 11, 12, 13, 15, and 16.

Montgomery, C. W., and Dathe, D., 1991. *Earth: Then and Now*. New York: Wm. C. Brown. Chapters 4, 7, 8, 10, 13, and 14.
Rhodes, F. H., and Stone, R. O., 1981. *Language of the Earth*. New York: Pergamon Press.
Weiner, J., 1986. *Planet Earth*. New York: Bantam Books.

Chapter 1: Those Creepy Continents. Creeping along to Antonin Dvořák's *New World Symphony* or Paul Winter's "Appalachian Morning."

Dewey, J. F., 1972. "Plate Tectonics." *Scientific American*, May, 56–68.
Hallan, A., 1983. *Great Geological Controversies*. Oxford: Oxford University Press.
Hurley, P. M., 1968. "The Confirmation of Continental Drift." *Scientific American*, April, 52–64.
Marvin, U. B., 1973. *Continental Drift: The Evolution of a Concept*. Washington, DC: Smithsonian Institution Press.
Menard, H. W., 1986. *The Oceans of Truth: A Personal History of Global Tectonics*. Princeton, NJ: Princeton University Press.
Paton, T. R., 1986. *Perspectives on a Dynamic Earth*. Boston: Allen and Unwin.
Wegener, A., 1924. *The Origin of Continents and Oceans*. London: Methuen.
Wilson, J. T., ed., 1976. *Continents Adrift and Continents Aground*. San Francisco: W. H. Freeman.

Chapter 2: The Fire Within. Exploding to Igor Stravinski's *The Firebird*.

Cass, R., and Wright, J., 1990. *Volcanic Successions: Ancient and Modern*. New York: Allen and Unwin.
Decker, R., and Decker, B., 1989. *Volcanoes*. New York: W. H. Freeman.
Fisher, R. V., and Schminke, H.-U., 1984. *Pyroclastic Rocks*. Berlin: Springer-Verlag.
Foxworthy, B. L., and Hill, M., 1982. *Volcanic Eruptions of 1980 at Mount St. Helens: The First 100 Days*. U.S. Geological Survey Professional Paper 1249, Washington, DC.
Grove, N., 1992. "Volcanoes: Crucibles of Creation." *National Geographic*, December, 5–42.

Harris, S. L., 1980. *Fire and Ice: The Cascade Volcanoes*. Seattle: The Mountaineers.
Kraft, M., and Kraft, K., 1975. *Volcano*. New York: Henry A. Abrams.
Macdonald, G. A., 1983. *Volcanoes*, 2nd edition. Englewood Cliffs, NJ: Prentice-Hall.
Macdonald, G. A., and Abbot, A. F., 1979. *Volcanoes in the Sea: The Geology of Hawaii*. Honolulu: The University Press of Hawaii.
Palmer, L., 1980. *Mt. St. Helens: The Volcano Explodes*. Seattle: Caroline House Publishers.
Rittman, A., 1976. *Volcanoes*. London: Orbis.
Simkin, T., and Fiske, R., 1983. *Krakatau 1883: The Volcanic Eruption and Its Effects*. Washington, DC: Smithsonian Institution.
Simkin, T., and Luhr, S., eds., 1994. *Paricutin: The Volcano Born in a Cornfield*. Phoenix, AZ: Geoscience Press.
Williams, H., 1963. *Crater Lake: The Story of Its Origin*. Berkeley: University of California Press.
Williams, H., and McBirney, A. R., 1979. *Volcanology*. San Francisco: Freeman, Cooper.

Chapter 3: Living on an Eggshell. Shaking along to any of the following: Bill Haley's "Shake, Rattle, and Roll," Momma Cass's "They Tell Me the Fault Line Runs Right through Here," or Carol King's "I Feel the Earth Move under My Feet."

Bolt, B. A., 1994. *Earthquakes*. New York: W. H. Freeman.
Gibbs, N., 1994. "Aftershock." *Time*, January 31, 26–39.
Gribben, J., 1978. *This Shaking Earth*. New York: Putnam and Sons.
Harris, J. B., and Kiefer, J. D., 1994. "Update on the New Madrid Seismic Zone." *Geotimes*, July, 14–18.
Nash, J. M., 1994. "The Next Big One," *Time*, January 31, 45–51.
Ritchie, D., 1984. *Superquake*. New York: Crown.
Walker, B., 1982. *Planet Earth, Earthquakes*. New York: Time-Life Books.

Chapter 4: And Away We Go. Sliding to Simon and Garfunkel's "Slip Slidin' Away."

Bolt, B. A., Horn, W. L., Macdonald, G. A., and Scott, R. F., 1977. *Geological Hazards*, 148–196. New York: Springer-Verlag.

Crandell, D. R., 1971. *Postglacial Lahars from Mount Rainer Volcano*. Washington, DC.: U.S. Geological Survey Professional Paper 677.

Crozier, M. J., 1986. *Landslides: Causes, Consequences and Environments*. Dover, DE: Croon Helm.

Cruden, M. J., and Krahn, J., 1978. "Frank Rockslide, Alberta, Canada." In B. Voight, ed., *Rockslides and Avalanches I: Natural Phenomena*, 97–112. Amsterdam: Elsevier Scientific Publishing.

Fisher, R. V., and Schmincke, H.-U., 1984. *Pyroclastic Rocks*, 297–311. New York: Springer Verlag.

McSaveney, M. J., 1978. "Sherman Glacier Rock Avalanche, Alaska, U.S.A." In B. Voight, ed., *Rockslides and Avalanches I: Natural Phenomena*, 197–256. Amsterdam: Elsevier Scientific Publishing.

Skinner, B. J., and Porter, S. C., 1989. *The Dynamic Earth*, 176–194. New York: John Wiley and Sons.

Voight, B., ed., 1978. *Rockslides and Avalanches I: Natural Phenomena*. Amsterdam: Elsevier Scientific Publishing.

Willams, H., and McBirney, A. R., 1979. *Volcanology*, 171–178. San Francisco: Freeman, Cooper.

Chapter 5: When the Waters Dance. Spouting off to Handel's *Water Music*.

Browne, P. R. L., and Lloyd, E. F., 1986. "Water Dominated Geothermal Systems and Associated Mineralization." In B. F. Houghton and S. D. Weaver, eds., *Taupo Volcanic Zone*, 145–212. New Zealand Geological Survey, tour guide A2.

Macdonald, G. A., 1983. *Volcanoes*, 323–344. Englewood Cliffs, NJ: Prentice-Hall.

Nash, J. M., 1993. "How Did Life Begin." *Time*, October 11, 66–74.

Rona, P. A., 1992. "Deep-Sea Geysers of the Atlantic." *National Geographic*, October, 105–110.

Shock, E. L., 1994. "Hydrothermal Systems and the Emergence of Life." *Geotimes*, March 12–15.

Williams, H., and McBirney, A. R., 1979. *Volcanology*, 319–346. San Francisco: Freeman, Cooper.

Chapter 6: Harvesting the Earth. Digging it with the Beatles' "Lucy in the Sky with Diamonds."

Argenzio, V., 1974. *Diamonds Eternal*. New York: David McKay.
Berton, P., 1958. *The Klondike Fever*. New York: Alfred A. Knopf.
Bruton, E., 1971. *Diamonds*. London: Chilton Book.
Cleary, D., 1990. *Anatomy of the Amazon Gold Rush*. Ames, IA: University of Iowa Press.
Craig, J., Vaughan, D. J., and Skinner, B. J., 1988. *Resources of the Earth*. Englewood Cliffs, NJ: Prentice Hall.
Epstein, E. J., 1982. *The Rise and Fall of Diamonds*. New York: Simon and Schuster.
Kesrah, C., and Craig, J. R., 1993. "The Importance of Observations in Geology with Reference to Gold-Nugget Formation." *Journal of Geological Education*, 41, 23–27.
Kessler, S., 1994. *Mineral Resources, Economics and the Environment*. New York: Macmillian College Publishing.
Mineral Commodity Summary, 1994. Washington DC: U.S. Bureau of Mines.
Mineral Information Institute (Update), 11, 5.
Morgan, M., 1967. *One Man's Gold Rush*. Seattle: University of Washington Press.
Ray, D. L., 1990. *Trashing the Planet*. Washington, DC: Regency Gateway.

Chapter 7: Winter's Breath. Flowing and grinding to Ralph Vaughan Williams's *Antarctic Symphony*, Warren Nelson's "Glacier Music," or Paul Winter's "Antarctica."

Bader, H., 1939, *Der Schree Und Seine Metamorphim*, Betroges Zur Geologic der Schwerz.
Charlton, W. 1983. *Ice Ages*. Alexandria, VA: Time-Life Books.
Chernicoff, S., and Venkatakrishnan, R., 1995. *Geology*, 467–502. New York: Worth.
Hambrey, M., and Alean, J., 1992. *Glaciers*. Oxford: Cambridge University Press.
Sharp, R., 1988. *Living Ice: Understanding Glaciers and Glaciation*. Oxford: Cambridge University Press.

Chapter 8: Life on the Edge. Riding the surf to "Oceans" by Second Chance or "Play of Waves" or "Dialogue of the Wind and the Sea," both by Claude Debussy.

Brown, R., ed., 1994. *State of the World*. New York: W. W. Norton.
Carson, R., 1961. *The Sea Around Us*, revised edition. New York: New American Library.
Duxbury, A., and Duxbury, A., 1994. *World's Oceans*, 4th edition, 23–307. New York: Wm. C. Brown.
Engel, L., and editors, 1969. *The Sea* (Life Nature Library). New York: Time-Life Books.
Garner, H. F., 1974. *The Origin of Landscapes*, 536–584. London: Oxford University Press.
Gore, R., 1993. "Andrew Aftermath." *National Geographic*, April, 2–37.
Groen, P., 1967. *Waters of the Sea*. London: D. Van Nostrand.
Groves, D., 1989. *The Oceans: A Book of Questions and Answers*. New York: John Wiley and Sons.
Leatherman, S. P., 1979. *Barrier Islands*. New York: Academic Press.
Mysak, L. A., and Lin, C. A., 1991. "The Tempering Seas." In C. Mungall and D. McLaren, eds., *Planet under Stress*, 134–148. Toronto: Oxford University Press.
Russell, R., 1967. *River Plains and Sea Coasts*. Berkeley: University of California Press.
Skinner, B. J., and Porter, S. C., 1989. *The Dynamic Earth*, 175–197. New York: John Wiley and Sons.
Stowe, K., 1987. *Essentials of Ocean Science*, 82–142. New York: John Wiley and Sons.
Wendt, H., 1969. *The Romance of Water*. New York: Hill and Wang.
Wenzel, H., 1979. *The Eternal Sea*. London: Abeland-Schumann.

Chapter 9: Summer's Fire. Broiling and drifting along to "Drought and Devastation" by Virgil Thompson or "Under the Sun" by Paul Winter.

Bonnifield, M. P., 1979. *The Dust Bowl*. Albuquerque: University of New Mexico Press.
Bryson, R. A., and Murray, T. J., 1977. *Climates of Hunger*. Madison: University of Wisconsin Press.

Cloudsley-Thompson, J. L., 1984. "Introduction." In J. L. Cloudsley-Thompson, ed., *Sahara Desert*, 1–17. Oxford: Pergamon Press.

Cooke, R., Warren, A., and Goudie, A., 1993. *Desert Geomorphology*. London: University College Press.

Garner, H. F., 1974. *The Origin of Landscapes*, 310–377. New York: Oxford University Press.

Greely, R., and Iverson, J., 1985. *Wind as a Geological Process*. Cambridge: Cambridge University Press.

Hollon, W. E., 1966. *The Great American Desert, Then and Now*, 9–21, 160–238. New York: Oxford University Press.

Hurt, R. D., 1981. *The Dust Bowl: An Agricultural and Social History*, 1–87. Chicago: Nelson Hall.

Jaeger, E., 1967. *The North American Deserts*, 1–32, 123–141. Stanford, CA: Stanford University Press.

Parfit, M., 1989. "You Could See It A-Comin." *Smithsonian*, June, 44–62.

Skinner, B. J., and Porter, S. C., 1989. *The Dynamic Earth*, 241–261. New York: John Wiley and Sons.

Stewert, J., and Tiessen, H., 1991. "Grasslands into Deserts?" In C. Mungalt and D. J. McLaren, eds., *Planet Under Stress*, 188–206. Oxford: Oxford University Press.

Chapter 10: Leaping Lizards and Cosmic Nights. Lights out as the earth plays *The Four Seasons* by Vivaldi, "And the Earth Spins" by Paul Winter, or "The Burgess Shale" by Rand Steiger.

Ager, D., 1993. *The New Catastrophism*. New York: Cambridge University Press.

Alluan, W. F., 1993. "The Dinosaur Hunter." *U.S. News & World Report*, June 7, 62–73.

Alvarez, L., Alvarez, W., Asaro, F., and Michel, H. V., 1980. "Extraterrestrial Cause for the Cretaceous-Tertiary Extinction." *Science*, 208, 095–1108.

Bakker, R. T., 1989. *The Dinosaur Heresies*. New York: William Morrow.

Gore, R., 1993. "Dinosaurs." *National Geographic*, January, 2–55.

Kauffman, E. G., and Erwin, D. R., 1995. "Surviving Mass Extinctions." *Geotimes*, March, 14–17.

Lemonick, M. D., 1993. "Rewriting the Book on Dinosaurs." *Time*, April 26, 42–49.

Monastersky, R., 1993. "Oxygen-Extinction Theory Draws Counter Fire." *Science News*, 144, 294.
Officer, C. B., 1990. "Extinctions, Iridium, and Shocked Minerals Associated with the Cretaceous-Tertiary Transition." *Journal of Geological Education*, 38, 402–414.
Stanley, S. M., 1987. *Extinction*. New York: Scientific American Books.
Toon, O. B., and Zahule, K., 1995. "All Impacts Great and Small." *Geotimes*, March, 21–23.
Watson, T., 1994. "The Crash." *U.S. News & World Report*, August 1, 54–63.

Index